P9-DOA-031

TO Kathy

The pleasure of working with you has been exceeded only by your performance!

With warm regards,

Will Raydor

September 16, 2005

Operational Performance Measurement

Increasing Total Productivity

"This book will add measurably to the success of those businesses and managers who read it and apply it."

John L. Mariotti, President and CEO
The Enterprise Group

"Imagine a pilot trying to fly an airplane with all its instruments blacked out. How would he navigate? In Operational Performance Measurement: Increasing Total Productivity, *Will Kaydos shows you how to build the instruments you need to keep your organization running efficiently and headed in the right direction."*

Bruce Sheridan
General Electric Mortgage Insurance Corp.

"Operational performance measurement works! In ten months, our performance measurement system enabled us to reduce defects 35%, increase productivity 15%, cut order cycle time by one-third, and increase sales at least 10%."

Jim Dickson, President
Consolidated Engravers

"Will Kaydos speaks to the reader as manager-to-manager. He blends just enough theory with sound advice and actual cases to clarify a very complex subject. Operational Performance Measurement: Increasing Total Productivity *is a practical handbook for the manager who wants to achieve real world results. I recommend it very highly."*

J. V. Gilmour, Corporate Controller
Harriet & Henderson Yarns, Inc.

Information about the author is on page xv.

Operational Performance Measurement

Increasing Total Productivity

by
Will Kaydos

S^{t}L

St. Lucie Press
Boca Raton London New York Washington, D.C.

Library of Congress Cataloging-in-Publication Data

Kaydos, W. J. (Wilfred J.)
    Operational performance measurement : incresing total productivity / Will Kaydos.
      p. cm.
    Includes bibliographical references and index.
    ISBN 1-57444-099-3 (alk. paper)
    1. Industrial productivity—Measurement. I.Title.
HD56.25.K39 1998
658.51'5'0287—dc21

                                 98-19227
                                    CIP

This book contains information obtained from authentic and highly regarded sources. Reprinted material is quoted with permission, and sources are indicated. A wide variety of references are listed. Reasonable efforts have been made to publish reliable data and information, but the author and the publisher cannot assume responsibility for the validity of all materials or for the consequences of their use.

Neither this book nor any part may be reproduced or transmitted in any form or by any means, electronic or mechanical, including photocopying, microfilming, and recording, or by any information storage or retrieval system, without prior permission in writing from the publisher.

The consent of CRC Press LLC does not extend to copying for general distribution, for promotion, for creating new works, or for resale. Specific permission must be obtained in writing from CRC Press LLC for such copying.

Direct all inquiries to CRC Press LLC, 2000 N.W. Corporate Blvd., Boca Raton, Florida 33431.

Trademark Notice: Product or corporate names may be trademarks or registered trademarks, and are used only for identification and explanation, without intent to infringe.

© 1999 by CRC Press LLC
St. Lucie Press is an imprint of CRC Press LLC

No claim to original U.S. Government works
International Standard Book Number 1-57444-099-3
Library of Congress Card Number 98-19227
Printed in the United States of America   4 5 6 7 8 9 0
Printed on acid-free paper

CONTENTS

PREFACE

Most managers wouldn't even think about getting on an airplane that was missing most of its instruments. But in business, these same managers are flying by the seat of their pants because they don't have relevant and reliable performance measures. Without performance measures, they can't really understand how their business processes work, the problems within them, and whether their attempts to improve performance worked as planned. That's why getting good performance measures in place is the first step a company should take on the path to improving quality, productivity, sales, and profits.

Performance measures provide many other benefits as well. These benefits are not commonly appreciated because most managers have never had the luxury of having reliable and comprehensive performance measures. Putting it another way, it's hard to miss what you've never had in the first place.

Two of the most important and least recognized benefits of performance measures are motivating an organization and helping to change its culture. By clarifying responsibilities, enabling establishment of specific goals, and providing positive feedback on accomplishments, performance measures are a catalyst for creating a culture of excellence, teamwork, and continuous improvement. If you don't believe this, just imagine being a football coach trying to develop a championship team under the following conditions:

- There are no yard-lines or goalposts on the field, so the team never knows where it is and how many yards it needs to score.
- The players know what position they are playing, but neither offensive nor defensive plays are called. Everyone just does what he thinks is best.
- The coach can only see five minutes of every game in fifteen-second segments, so the feedback he can give the team is quite

limited. The only other reliable information he gets about his team's performance is the total yards gained by each side and the final score — but this is received two days after the game is played.

You wouldn't find many people willing to coach under those conditions — but most managers are working with similar handicaps every day. They don't have clearly defined objectives, don't understand what to do to support company strategy, and don't get useful feedback about their performance so they can focus their resources where they are most needed.

Instead, upper level managers are assigned financial objectives that have little relationship to strategy and receive accounting reports that only indicate fairly large deviations in performance several weeks or months after something in the business has changed. Middle managers may get similar information, but it is too symptomatic and too late to be of much practical value for either monitoring or improving performance.

Most companies supplement their accounting reports with some non-financial (operational) measures to fill a few holes in the performance puzzle, but not all of them. At best, these operational measures can indicate some problems earlier than accounting reports, but they seldom go much beyond saying a problem exists. At worst, the operational measures may be misleading, because they don't reflect what customers want, aren't linked to a strategy, do not make up an integrated system, and provide only a partial picture of performance that reflects what is easy to measure, not what is important to measure.

Surveys I have conducted of middle and executive level managers in a broad range of industries indicate that nearly every manager wants good, fair performance measures, but virtually no one has them. Managers are apparently keenly aware of their need for performance measures, but this has not resulted in action to fill the need. Granted, a small number of companies have good performance measures throughout their business but that is the exception, not the rule. In fact, one of the distinguishing features of the Baldrige Award winning companies (and their peers at the state level) is the depth and breadth of their performance measures and how these measures reflect customer and company values.

Knowing the score is important, but what really matters in business is improving the score. That's what this book is about — explaining how to design and develop performance measurement systems that will significantly increase productivity, quality, sales, and profits. This is accomplished by first presenting a general theory of performance measurement consisting of:

- A measurement model which can be applied to *any* business activity.
- Technical and cultural requirements which must be satisfied for a measurement system to be truly effective in improving performance.

The general theory is important because it provides a structured framework for developing performance measurement systems. Without this framework, those who venture into the uncharted territory of developing a measurement system are likely to fall into a number of pitfalls that could be avoided.

Like any scientific theory, the principles and procedures given here cannot be proven, but they have been validated by applying them in a variety of business situations. The resulting measurement systems have been instrumental in producing substantial improvements in quality, productivity, costs, and sales.

In order to enable the reader to achieve similar results, procedures for applying the theory to the real world are provided. Examples and case studies are used to illustrate how these procedures are used, the performance measures that were developed, and the results obtained.

Performance measurement is powerful stuff. It is not the answer to every performance problem, but it is a fundamental discipline that belongs in every manager's toolkit. This new age of low-cost computers, flexible database management software, and data collection devices makes it possible for companies to have performance measurement systems that wouldn't have been economically feasible ten years ago.

I believe that the day will come when comprehensive performance measurement systems are found in the majority of companies and that performance measurement is commonly taught in business schools. There are just too many advantages for companies, managers, and employees for this not to happen. How long this will take is anyone's guess, but if you apply the principles and techniques provided in this book, you can start reaping the rewards instead of waiting to follow someone else.

Will Kaydos

THE AUTHOR

In his thirty years of experience as a manager, executive, and consultant, Will Kaydos has developed performance measurement systems that have led to significant improvements in quality, productivity, sales, and profits. Prior to founding The Decision Group, he was senior vice president of manufacturing and of finance for a $160,000,000 corporation, where he used performance measures to more than double manufacturing productivity and substantially reduce overhead costs.

He is the author of *Measuring, Managing, and Maximizing Performance*, which has been called one of the best books about quality improvement by managers, quality professionals, and educators alike. Will is also the author of *Exceeding Expectations* and of numerous articles and commentary on performance measurement and quality improvement which have appeared in *Quality Digest, Business Horizons, Quality Progress*, and other periodicals.

An internationally recognized authority on performance measurement and quality improvement, he is a frequent speaker for universities and professional associations. Will has an MBA from the Wharton School and a BS from Clarkson University.

He can be contacted at: The Decision Group, PO Box 15005, Charlotte, NC 28211; (704) 364-4619; http://www.decisiongroup.com.

DEDICATION

To Susan, the light and love of my life

1

WHY MEASURE PERFORMANCE?

Performance measures provide managers, front-line employees, and companies with a broad assortment of both cultural and technical benefits. These benefits go far beyond the bottom line, but they are not commonly recognized. While it is not a prerequisite to implementing performance measures, an understanding of these benefits will give managers insight into what makes a good measurement system and how performance measures should be used.

BENEFITS FOR MANAGERS

Improved Control

Feedback is essential for control of any system. Without the feedback provided by sight, sound, and touch, we humans would not be able to identify threatening or favorable situations and there would undoubtedly be fewer of us on the planet. The same is true for companies. When managers don't have timely and meaningful feedback, companies fail to recognize opportunities and become much more vulnerable to hazards that can threaten their existence.

The feedback provided by performance measures gives managers better control over their areas of responsibility, whether it is a department, a plant, or a division. With measures in place, deviations in performance are detected earlier, enabling managers to step in and minimize the damage or make the most of the opportunity. Performance measures also prevent managers from getting blind-sided with bad news. If you have ever experienced some unpleasant surprises at the end of a month or quarter, it indicates you don't have appropriate measures in place. With adequate measures, the news may not always be good, but at least it won't be a surprise.

Clear Responsibilities and Objectives

Good performance measures clarify who is responsible for specific results or problems. They specify what "good performance" means for each person, manager, or operating unit in unmistakable terms. This has several distinct advantages for managers:

1. Everyone knows what they are supposed to accomplish. There is no question about what matters most, so when daily operating decisions have to be made, it is more likely the right course will be taken.
2. Everyone knows how well they are performing. This creates a self-correcting feedback loop that reduces negative deviations in performance.
3. Everyone becomes accountable for only *their* performance and not problems created by someone else. Anyone who has ever had their performance questioned because someone else didn't do their job properly, knows how discouraging that can be. Good performance measures make it clear who owns what parts of a performance problem. By doing so, they dramatically reduce inter-departmental finger-pointing and buck-passing.

Strategic Alignment of Objectives

Performance measures are probably the best way to communicate a company's strategy throughout an organization. Of course, this means a company must develop a strategy and determine what each operating unit must accomplish to execute it. This requires establishing a company's strategic objectives and then breaking them down into lower level objectives and corresponding performance measures. When a company's performance measures reflect its strategy, they assure everyone is working toward the same objectives and not going off in different directions.

Performance measures are also essential for assessing the effectiveness of a strategy. Unless a company's key business processes are under control and meeting their defined performance objectives, there is no way to tell whether a strategy is effective or not.

Understanding Business Processes

"Measurement is the first step that leads to control and eventually to improvement. If you can't measure something, you can't understand it. If you can't understand it, you can't control it. If you can't control it, you can't improve it."

H. James Harrington

When it comes to understanding a production process, the simple fact is that if you aren't measuring a process, you cannot understand how it works. You may know what goes into a process and what comes out the other end, but understanding how it works means knowing what happens in the middle, what factors affect its performance, how it will behave if something in the process changes, and what the process is capable of doing.

When performance measures are not in place, there is typically a big difference between how managers think a process works and the way it actually works. Here are a few examples that indicate how large that gap can be.

- A CEO believed rejects in a complex manufacturing operation were running about 10% — which is certainly not something to brag about. When measures were put in place, they showed the true reject rate was over 30%. The CEO is still recovering from shock.
- Everyone believed that on-time delivery performance in a service company was 90% or better. Measures showed that to be true for a select group of customers, but for most customers, it was closer to 60%.
- In a manufacturing plant with only twenty employees, what was believed by everyone to be the number one cause of production downtime turned out to be sixth on the list. What was more surprising, was that the two largest problems were never mentioned during preliminary surveys of the staff to determine the possible causes of downtime. As close as everyone was to the problems in the plant, it is difficult to believe their judgment could be so wrong. However, this illustrates just how unreliable subjective judgment of performance problems can be.

Similar examples would be easy to find in most companies. I have never encountered a situation where newly implemented performance measures did not show a substantial difference between management's perceptions of what was happening and what was actually happening. This should not be surprising. No manager can be everywhere all the time. Even if this was possible, storing and assimilating all the performance data produced by the typical business process is far beyond the mental capacity of anyone.

There is a very big difference between understanding *some* of the problems in a process and understanding *all* of them. This is certainly one of the primary reasons so many performance improvement initiatives fail. A case in point was a company that spent $180,000 to replace a piece of equipment that the managers had identified as being the primary bottleneck in the whole process. They had assured headquarters replacing this one piece of equipment would greatly increase productivity and output. Unfortunately, after the $180,000 was spent, there was no noticeable change in either output or productivity.

Needless to say, no one was pleased with the results, but nobody could explain why performance had not improved. The mystery was solved when performance measures were implemented in the plant. They showed that the "big bottleneck" was not the only problem that needed to be addressed. There were six other major causes of waste, downtime, and poor quality that had to reduced for the process to reach the productivity objectives.

Why had such a serious error in judgment been made? The reason was that the piece of equipment in question frequently broke down for short periods. Like being repeatedly stabbed with a pin, the aggravation of the frequent breakdowns was much greater than the actual damage. What seemed like 90% of the problem was really only 10%. Another case of human perception being quite different from reality.

Knowing What a Process Can Do — Its Capability

Understanding a process also means knowing its capability — the limits of what it can do. At any point in time, every process, mechanism, organization, or person, has a capability. By definition, this capability cannot be exceeded, even though one of the enduring cliches of the sports world is "giving 110%."

Knowing the capability of a machine is essential for determining what action needs to be taken to correct a problem. If a machine is not producing good products but has the capability to do so, something must be wrong with the machine and it must be repaired. On the other hand, if it does not have the capability to produce good products, "repairing" the machine will not solve the problem because nothing is broken. The only way insufficient capability can be corrected in a machine is to improve its capability. This might be accomplished by installing better parts, a more powerful motor, or even replacing the machine.

"If it ain't broke, don't fix it."
Anonymous

Managers should take this old cliché seriously, because trying to fix what isn't broken is an exercise in futility. There are many cases where just getting a process to work as it should, made dramatic improvements in performance. However, quality authorities say that 80–90% of quality problems are the result of processes that are not capable of doing what is being demanded of them. Measurement is the only way the capability of a machine or a process can be determined. That is one reason the sports pages are full of statistics. They are used by fans, writers, and anyone inclined to bet on the outcome of a game, to assess the capability of competing teams and players.

When processes and their capabilities are not understood, it is hardly surprising that so many attempts to improve performance fail. Surveys of companies that have implemented such concepts as material requirements planning, total quality management, and reengineering, indicate that only 25 to 35 percent of them achieved meaningful results.* "Reengineering" a process makes sense if you know the process is operating properly and it cannot do what is required. But if that hasn't been determined, then introducing radical changes may be totally unnecessary and could make matters worse. On the other hand, applying a Band-Aid when major changes are needed doesn't work either.

No doubt factors such as the lack of top management leadership and a poor understanding of the concepts account for many failures. But it is also clear managers are sometimes applying solutions to problems that don't exist and failing to recognize many problems that do exist.

Improved Quality and Productivity

As shown by Figure 1-1, quality improvement is simple in principle. It is just a matter of determining the difference between actual and desired performance and then changing or improving the production process to improve its performance.

---

* (Please see "Endnotes" at end of this book.)

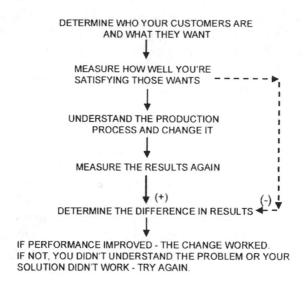

DETERMINE WHO YOUR CUSTOMERS ARE
AND WHAT THEY WANT

↓

MEASURE HOW WELL YOU'RE
SATISFYING THOSE WANTS

UNDERSTAND THE PRODUCTION
PROCESS AND CHANGE IT

↓

MEASURE THE RESULTS AGAIN

↓ (+)

DETERMINE THE DIFFERENCE IN RESULTS

↓

IF PERFORMANCE IMPROVED - THE CHANGE WORKED.
IF NOT, YOU DIDN'T UNDERSTAND THE PROBLEM OR YOUR
SOLUTION DIDN'T WORK - TRY AGAIN.

Figure 1-1   The Quality Improvement Process

However, to make quality improvement work, the following measurements are needed.

1. **The size of the gap between what customers want and what they are getting.** This determines the size of the performance problem. In fact, without knowing this gap, you don't even know if there is a problem.

2. **Measurements of the process providing the goods or services.** This provides an understanding of the process. Without measures, how do you know where the problems are located and which ones should be attacked first? The answer is you can't know, which is why getting performance measures in place should be one of the first steps anyone takes when trying to improve quality and productivity.

3. **The size of the performance gap after changes have been made to improve the process** providing the goods or services. This tells you whether your attempts to improve performance worked. Without this measure, you are just guessing or engaging in wishful thinking.

Trying to improve performance without these measures is like shooting at targets in the dark. You will hit a few, but your score won't be anything to brag about because you won't know where the targets are or how to adjust your aim.

In almost all cases, implementing performance measures will, by itself, noticeably improve performance. In my experience, introducing measures typically improves quality 10 to 20% in a few weeks. Why the improvement? Certainly not because the measures were used to punish or reprimand anyone. Instead, the most important reasons for this initial improvement in performance are:

- The measures define what is important and focus attention on those issues.
- Developing measures requires establishing and communicating quality standards, which are usually not understood by everyone. It is amazing how often something is being done the wrong way because the person has never been told what "the right way" is.
- When performance becomes visible, no one wants to look bad to their peers or themselves. In one noteworthy case, when a group saw how poorly they were performing relative to other companies, they were so embarrassed they immediately took it upon themselves to assign teams and individuals to attack the problems. No one in management said a word about their performance, but within a few weeks rejects and rework in the department decreased 40%.

These initial gains can only be expected when the performance measures meet the criteria defined in subsequent chapters. Poorly conceived and implemented measures are as likely to cause performance to decrease as to increase.

More Efficient Allocation of Resources

Managers can be aware of most of the problems their work group experiences, but without measures in place, there is no way for them to know the relative importance of those problems. When unable to differentiate between the dozens of operating problems that are found in most business processes, managers have the following options.

1. **Ignore the whole situation.** This is probably the most common reaction when it seems there are too many problems to address and what's important cannot be separated from what's not.
   Result: Aggravation and frustration, but no action.

2. **Have a meeting and tell everyone they must do better.** Unfortunately, no one is told specifically what he or she needs to do better, so everyone assumes the other person is responsible.
   Result: Nothing happens except finger-pointing and rationalizing that "I'm really not responsible."

3. **Try to solve all the problems at the same time.** This spreads the organization's resources so thin, efforts get focused on treating symptoms and making temporary patches instead of making lasting improvements to processes. Management may also lose control of the situation and some resources will be wasted on areas of little return.
   Result: Boxes of bandages get applied and a few aspirin are taken, when some parts of the patient really need surgery. Many people will also be discouraged by seeing resources wasted on trivial issues and not achieving meaningful results.

4. **React to whatever was most important last week.** Unfortunately, next week may present a completely different picture, forcing another realignment of priorities.
   Result: Considerable confusion and wasted effort. As in (3) above, few permanent solutions will be implemented.

One of a manager's primary functions is to know where to apply resources for the greatest return. Since the demand for a company's or department's resources always exceeds the supply, knowing where to get the most benefits for the least investment is very beneficial to a manager's peace of mind — as well as to a company's profits.

Good measures greatly improve managers' decisions about where to allocate resources by establishing the relative importance of problems and opportunities. Even if the measures of potential or loss are not very accurate, they will still differentiate between major and minor issues. Managers will still have to assess the investment required, the probability of success, the availability of needed resources, and other factors to determine which opportunities to pursue, but the information provided by performance measures will greatly narrow the field.

Working on the right things is important at all levels of a company, not just at the top. Although the merits of problem solving and process teams are well established, their efforts should always be focused on important issues. One school of thought is that teams should be allowed to determine their own projects. This may be appropriate in some cases, but I believe a more effective approach is for a steering committee to select projects for teams that are derived from performance measures.

An example of what can happen when a team has no factual basis for determining which issues to address was provided by a former team member in a company that supposedly had a successful quality improvement program. In a discussion about the company's program, I asked her how she enjoyed the team experience. Her reply was:

> *"It was terrible. All we accomplished in twelve months was debating what the team should work on. Everyone had a different idea, which seemed to be based on what would make their job easier, not what was best for the company. We never did agree on what to do. It was the most frustrating and aggravating experience I've ever had."*

This may be an extreme case, but it points out what can happen with no direction or objective information about problems and opportunities. It illustrates why performance measures are such an important part of any quality improvement program. After all, if you don't know where you are, how can you determine which path to take? Empowering teams or workgroups to take action without direction or giving them a performance measurement compass that will enable them to choose meaningful projects can lead to discouragement as well as poor results.

Bob Reber, president of Local 1375 of the United Steelworkers of America has worked with several companies introducing participative programs. He sees it this way: "Empowered employees need and want to know how to most productively focus their mental horsepower. They don't want to go off aimlessly any more than you or I do. They want and need guidance on how to meaningfully contribute."[2]

By implementing performance measures and focusing improvement efforts on the major problems and opportunities they reveal, tangible results will almost always be achieved within several months. Nothing is a more powerful motivator than a little success. It demonstrates performance can be improved, that ordinary people can do it, and that it doesn't take forever to make a difference in performance. Those directly involved in improving performance can see what they have accomplished and are encouraged to do more. When they see results, many of the managers and employees standing on the sidelines will become believers and want to participate.

The effect is like what happens when a professional sports team starts winning a few games after being in the pits for a while — suddenly everybody starts wearing their team jackets again. Everybody wants to be associated with a winning team, and no one wants it more than the team members themselves.

Better Planning and Forecasting

Managers who understand how processes work and their capabilities, are obviously able to make more reliable plans and forecasts than those who do not. Sometimes external events can make forecasts miss expectations, but when forecasts are based on a poor understanding of the dynamics of a process, they will be inherently flawed.

For example, when a company's sales failed to meet the forecast by a wide margin, analysis of the sales data showed there were performance measures that would have indicated the sales problem if they had been in place. These measures would also have provided enough lead-time to take corrective action to reduce the size of the predicted sales deficit. Later, when the measures were implemented, the wide swings in sales that had occurred in the past were drastically reduced. This not only increased sales, it reduced manufacturing costs and improved delivery performance as well.

Performance measures provide insight into how various operations will be affected by changes in inputs and both external and internal factors. It takes some time to acquire this understanding of process dynamics, but once gained, it will make operational plans and financial projections much more reliable.

The Freedom to Delegate

Managers are often reluctant to delegate because they are afraid things might get out of control without them knowing about it. This is perfectly rational behavior. After all, who wants to wake up and find they are going sixty miles an hour toward a brick wall that is only twenty feet away?

However, when managers can stay in touch with what is happening through performance measures, the fear of delegating disappears. So too, does the tendency to micro-manage, which is degrading to most employees. Being able to manage from a distance can do more to increase a manager's personal productivity and mental health than just about anything.

CYA and Defending your Position

In the business world, CYA (Covering Your Ankles) and arguing your position is much easier to do when you have the numbers on your side. Without hard numbers, you're just another person with an opinion — and everybody has one of those. This is especially true when dealing with a superior who has ingrained beliefs about problems, people, and the way things work.

In one such case, a general manager had formed the belief that all late shipments were caused by the satellite plants and that the managers of the plants were essentially incompetent. When the plant managers pointed out chronic raw material shortages, incorrect production schedules, and production demands that were impossible to meet, the general manager listened. However, when he brought the reported problems up with the responsible department heads, they were dismissed as "one-time exceptions." This only reconfirmed his belief that the plant managers were at fault. Since the plant managers didn't have any hard facts to support their arguments, the general manager dismissed the complaints as lame excuses.

However, when performance measures were implemented in manufacturing, purchasing, inventory control, and production planning, it didn't take long for it to become painfully clear to the general manager that the plant managers were correct in their assertions about the support departments' performance. When the managers of those departments were confronted with the facts, it took only a few months for them to make some remarkable improvements in their departments' performance.

Whether it is CYA about performance problems caused by others or arguing your position about some other matter, performance measures provide powerful ammunition. The other side of the coin, however, is that the measures will also expose some of your problems that you would rather keep hidden. However, I have never had any manager complain about making their own performance visible. Good managers don't mind being held accountable for what they control, which is why the overwhelming majority of them want performance measures.

Changing a Company's Culture

Performance measures by themselves won't change a company's culture, but they are a powerful catalyst for doing so. As will be explained, certain cultural conditions must be satisfied in order to successfully implement performance measures, but once measures are in place, they will influence an organization in several ways.

By defining common goals, performance measures promote teamwork. A team can't exist if there is no shared goal, so when group goals are defined, a clear reason for teamwork is created. This alone moves an organization in the right direction, much as having a common enemy unites people in time of war. Leaders must take over from there, but clear goals get the ball rolling.

Teamwork is also reinforced by defining what each team member must accomplish for the team to achieve its goal. In football, the quarterback or coach calls the play, which tells each individual what to do to maximize the team's chances of moving toward the goal. By establishing accountability and defining objectives, performance measures "call the play" for the sub-units of a company so they know what to do to support higher level objectives.

When clear objectives and responsibilities are defined, conflicts between departments will be reduced because there will be few "fuzzy" areas of responsibility and conflicting goals. Then, when initial measures show that everyone has room for improvement, finger-pointing and making up excuses will also be reduced. Performance measures also focus attention on problems instead of personalities, which further reduces conflict between individuals and work groups.

A more subtle change in culture is introduced when objective performance measures become available. Numbers get people oriented toward rational discussion instead of debating on the basis of feelings and opinions. Having solid information available changes the way an organization approaches problems. Once people start using measures to identify and solve problems, they will want to continue that way. Of course, some training in problem solving will facilitate this change in behavior.

A good performance measurement system keeps everyone honest by giving no one a place to hide. When performance becomes readily visible, it can make some people very uncomfortable, but if a company's leaders set the right tone, fear is quickly replaced with more open and honest communication. This has a positive impact on the interactions between employees and managers.

BENEFITS FOR EMPLOYEES

Clear Responsibilities and Objectives

Knowing what "good performance" means is just as important for employees as it is for managers. Like anyone, front-line employees are most concerned about the issues that are closest to their own interests. Consequently, while they are interested in how their company or division is doing, they are most interested in what *they* are supposed to accomplish, their own performance, and the performance of their work group.

As one railroad old-timer put it when talking about his company's quality improvement program:

> *"Our president gave us a big speech about improving quality and he kept saying we had to get better to get the business back from the trucking companies. I agree with that, but what am I supposed to do about it? I don't talk to people who ship things, he does. What can I do to get the business back? When someone tells me that, I'll do it. Until then, don't bother me with another bunch of words."*

In essence, this is much of what effective performance measurement is all about. Telling everyone in a company what is important for them to accomplish and giving them feedback on how well they are doing. Making people accountable is one thing, but specifically saying what they are accountable for is quite another.

In the case of our railroad man, if his manager could show him how certain aspects of his work could affect delivering freight with no damage and how that would make the railroad more competitive, I would be willing to make a sizable bet that he would try to improve his performance.

Seeing Accomplishments and Receiving Recognition

For managers and employees alike, it is difficult to take pride in accomplishments that can't be seen. Even people making physical products get very little opportunity to see what they have accomplished, because they only do a few steps of the total manufacturing process. Knowing how many service calls were completed on time, how much customer satisfaction increased, and last week's error rate is very important to the people doing the work. Unfortunately, the only time most employees get feedback is when something goes wrong.[3]

The capacity for managers to coach, appraise, give feedback, and reward performance is the area where employees say their managers need to improve the most.[4] Performance measures may not change a manager's behavior, but by providing the means to identify where feedback, coaching, and recognition is needed, they greatly facilitate the process. Achievement and recognition for contributions are two primary motivators of employees.[5] When their accomplishments become visible, their sense of achievement is reinforced and it becomes more likely that managers will recognize their contributions.

Being Evaluated Objectively

Fair reward and recognition is the cornerstone for building a motivated and effective organization. Even with good measures, judgments have to be made about the performance of individuals and groups, but without them, properly equating performance with reward is practically impossible. As one employee said: "Sometimes I don't like my performance being measured, but I would rather have my performance evaluation based on what I accomplish rather than someone's opinion."

It is not always practical to measure an individual's performance, nor is it required for performance measures to be effective in improving quality and productivity. However, measures of individual performance will make any favoritism more apparent, so performance measures act as a safeguard against favoritism and prejudice for both individuals and managers. Managers and employees alike are better served when performance appraisals are based on as much objective information as possible. Good performance measures can only improve the fairness of evaluations.

More Empowerment

As mentioned before, performance measures enable managers to delegate responsibility and manage from a distance. Performance measures also discourage micro-managing by focusing attention on results and away from the minute details of how they were obtained. This will lead to more freedom for supervisors, employees, and work teams, making work more enjoyable. Life also becomes more enjoyable for managers when they don't have to worry about being blindsided by problems, don't have to get into all the details of operations, and see the higher performance yielded by delegation and empowerment.

SUMMARY

Performance measures provide benefits to managers, employees, and their companies from a social as well as a technical perspective. These benefits are not very well understood because few companies have good measurement systems and it is difficult to understand or appreciate something you have never experienced.

Perhaps the greatest misunderstanding about performance measures is that they lower morale, because people don't like to have their performance measured. This can be true if the performance measures are used to find fault and punish individuals. But when they are used in a positive manner to identify problems and reward accomplishments, performance measures are a powerful tool for motivating an organization and changing its culture.

By letting managers delegate and spend their time on the problems and priorities that really require their attention, performance measures greatly increase the personal productivity of managers and employees alike.

2

WHAT IS MEASUREMENT?

Mention "measurement" to someone, and they will usually think of some physical unit of measure such as feet, degrees, or grams. That is one type of measurement, but just what is measurement and what other types of measurement are possible?

Measurement consists of assigning a numerical scale to the size, value, or other characteristic of a tangible or intangible object. The scale could be as simple as 0 to 1 (bad or good), 0 to 10 (as in athletic competitions), or a logarithmic scale like the Richter Scale used to measure the magnitude of earthquakes.

The first characteristic to note about measurement is that all measures are relative. A measure that is not referenced to something else has no meaning. For example, assume a sales representative's performance is measured by the percentage of prospects from a given group that make a purchase each month. For the first month he achieves a 53% score. What does this mean? Is this good or bad performance?

What if I tell you that six months ago, his score was 32%? The 53% figure now seems like a respectable score, doesn't it? But what if I now tell you that the average score for everyone else in the sales force is 88%? The 53% score that seemed good a few seconds ago, now looks pretty sorry, doesn't it? Clearly, without a reference for comparison, all measures are meaningless numbers. Consequently, if something is to be measured, one of the first considerations is what to use as a basis for comparison.

"Nothing is good or bad but by comparison."

Thomas Fuller

Measurement Standards

When the reference for comparison is an internationally recognized standard, such as grams, meters, seconds, or volts, the measurement will be called *standardized* measurement. Countable items such as dollars, defects, or late deliveries can also be considered standardized because everyone agrees what a given number represents. (This does not apply, of course, in the Land of Elastic Numbers, otherwise known as Washington, D.C.).

It should be noted that all of the international standards are purely arbitrary and have become accepted by custom and law. At one time, the standard for the yard was from the tip of the king's nose to the end of his outstretched hand. That supposedly worked quite well until the king died and was replaced by Shorty the Ninth, causing fortunes to be made and lost overnight. Eventually the English and metric systems of standards were developed to provide the stability necessary for trade. They are equally valid standards, although only a few backward countries like the United States still use the English system.

Where no accepted standard exists, the measure will be called a *relative* measure. A relative measure can be compared to itself at some other point in time or to the same measure in another system. For example, a customer satisfaction index will indicate a change in satisfaction with a company's products over time. If the products were toasters, a valid comparison could be made between different models, assuming the index was derived the same way at the same time. Trying to compare a company's satisfaction index with some other company's satisfaction index would be quite another matter, unless identical techniques and closely similar sample populations were used.

When using something other than a recognized standard as a basis for measurement, care must be taken to be sure any comparisons are valid. For example, a measured change in customer satisfaction would be questionable if the sample of customers used in the first survey had a much different income level than those in the second.

Measurement Methods

Once a reference for comparison has been established, how the measurement is going to be accomplished must be determined. If the object or condition itself is measured, the measurement is called *direct* measurement. Measuring the length of a board and counting rejected parts are direct measurements.

The other way something can be measured is to measure its effect rather than the item itself. This is called *indirect* measurement. For example, employee turnover and absenteeism can be indirect measures of morale. Even some physical qualities are measured indirectly. In a practical sense, using a meter to measure voltage is a direct measurement, but what is really being measured is the effect of the electricity, not the electricity itself.

Most indirect measures could more accurately be called *indicators*, because while they will show a change in a variable, they may not provide a reliable measure of the degree of change. For example, if a satisfaction index goes from 10 to 20, we can be quite sure customers are more satisfied, but it would not be correct to say they are now twice as satisfied as they were before. To avoid unnecessary complication, the term "measure" will be used for any number, including indicators, that relates to the size, quantity, or other property of anything.

Measurement Techniques

The possible combinations of measurement reference and measurement method are shown by Figure 2-1. The standardized-indirect combination is interesting, because at first glance, it seems to be an impossible combination. However, scientists commonly use indirect techniques to measure physical parameters with magnitudes that stretch the imagination.

Which technique should be used? That all depends on what you need to know. If you have a piece of pipe about fifty feet long and need to know if it is longer than another piece located 1000 miles away, a standardized measure is needed (although you could use a relative measure

	METHOD	
REFERENCE	**Direct**	**Indirect**
Standardized	Measures of physical parameters and countable items	Determining physical measures by effects — deriving a planet's weight from its effect on another's orbit
Relative	Measures derived from countable items — complaints/sale, defects/car, inventory turns	Measures of qualities and abstract attributes— satisfaction, morale, helpfulness, kindness, honesty

Figure 2-1 Measurement Techniques

by sending a very long fax!). However, if the pieces are about two feet long and in the same room, all you have to do is hold them side-by-side to determine which is longer.

The choice of the measurement technique to use should be based on the following factors:

1. What questions must be answered?
2. What techniques are feasible for producing the measure?
3. What is the most economical and reliable method of making the measurement?

Regarding the questions to be answered, the more important questions managers have to address are

- How are we doing — are we getting better or worse?
- What are the short and long-term trends in performance?
- Did the changes we made improve performance?
- What's not working as it should?
- Which departments, managers, or supervisors need some help?
- Where are the largest opportunities for improving performance?
- Where should we be applying our resources for the most return?
- Which departments need additional people? Which ones have a surplus?
- Where can we expect to be in the next 3 to 6 months?
- What are the bottlenecks or limiting factors in our key processes?
- Where do we need to make radical changes to the way we do things?

For these and most operating questions facing managers, the change and trend in the value of a variable or its value relative to other variables is all that is needed to provide the answer. For that, relative measures will do the job very well. Whether direct or indirect measurement is used depends on the variable, the process, and the economics of the situation. Direct measurement is possible for most technically oriented measures. Both direct and indirect measures are generally applicable for softer parameters such as morale, attitudes toward various issues, and satisfaction.

Highly accurate, standardized measures are simply not necessary for the practical application of performance measures in the typical business environment.

Measuring the Unmeasurable

Given the options available for measuring something, is it possible to measure anything? If the objective of measuring is to provide a precise, unequivocal number that will tell a manager everything he needs to know, the answer is "no." But if the objective is to provide reliable and meaningful information, which will improve the quality of managers' decisions, the answer is definitely "yes."

Measuring intangible concepts is far more common than most people appreciate. The "quality" or "appeal" of television programs is measured by the Nielsen index; various surveys and indices measure "attitudes" of organizations, and customer satisfaction with a variety of products and services is measured every day. Obviously, most of this information must be useful to someone or they wouldn't keep doing it. Even highly subjective factors like the quality of ice-skating and movies are routinely measured by knowledgeable judges in these fields. While someone may find fault with any particular instance, these measures are reliable guides once the user understands what they represent. For example, I find Siskel and Ebert's "thumbs up" measure of movie quality to be generally on the mark for my tastes. Someone else may not find that to be true for their tastes.

In a broad sense, "measurement" means assigning a number to a property of an object. In the simplest case, a measure could have two values, zero or one. The values assigned to a variable can be established by means such as counting, measuring with instruments, panels of judges, and other methods. In addition to the characteristics already described, performance measures can be put into the following categories:

- **Qualitative or subjective** — When numbers on a scale are assigned by human judgment. This does not necessarily imply there is any bias in the measure.
- **Quantitative or objective** — When measures are derived from physical measurements or countable units.
- **Attribute** — When a characteristic, such as a defect, is measured as either being present or not.
- **Variable or continuously variable** — When the degree or extent of a variable is measured on a continuous scale. The dimensions of a table top are variables; dents are attributes since they are counted as either being present or not.

It cannot be proven, but I believe anything can be measured to a useful degree, especially in a business environment. If something can't be measured directly, it must have an effect, which can be measured. If a process has no intended effect, it is clearly not worth measuring in the first place. More specifically, any production process can be measured if what it is supposed to accomplish and how it works are understood. As a minimum, every process must have at least one customer whose satisfaction can be determined. How far beyond this point measurement can be taken depends on the process and other factors.

What do you want to measure in your business? Customer satisfaction? Employee satisfaction with management? Administrative productivity and quality? Productivity of research and development? Inventory system performance? Performance of a sales organization? Competitive position of products? Service department performance? Scheduling effectiveness? Employee buy-in to management initiatives? These and other outcomes are currently being measured by companies on a routine basis. However, as will be explained, it is necessary to measure more than outcomes to improve performance.

The real question is not whether something can be measured, but whether it is worth the effort and money to do it. It may require some creative thinking and changes in the way things are done in order to acquire the necessary data, but these are not insurmountable barriers.

> *"Count what is countable,*
> *measure what is measurable,*
> *and what is not measurable,*
> *make measurable."*

Sound advice, but these words did not come from some modern management guru. Their source is Galileo Galilei (1564–1642), who was apparently ahead of his time in more ways than one.

SUMMARY

We usually think of measurement in terms of direct, standardized measures of physical properties such as length or weight. However, for the purpose of providing managers with information they can use to improve the timeliness and quality of their decisions, precise measurements are not required. For virtually all business decisions, direct and indirect relative measures are sufficient.

If a business activity is supposed to accomplish anything, its performance can be measured — providing how the process works and what it is supposed to accomplish are well understood. In the real world, however, the understanding of customer requirements and how processes work is often very superficial. That is why implementing performance measures is, to a large extent, a learning process.

Measuring performance in a business is not always easy. It can be challenging, may require some creativity, and will probably involve some trial and error, but it can always be done.

3

MEASURING PRODUCTION PROCESSES

THE BUILDING BLOCK OF PERFORMANCE MEASUREMENT

The building block for measuring performance is the "production process" shown by Figure 3-1. The entire production sequence consists of receiving work to be accomplished from a "vendor," adding value to the product (goods or services) with the production process, and passing the modified product on to a "customer," who may be the end user or just the next step in a chain of events.

Every business activity is a production process, with vendors and customers that may be unique or shared with other processes. Figure 3-2 shows how the production process model can be applied to some common business activities.

Since every business activity is a production process, a business can be viewed as a complex set of small (micro) processes linked to form larger (macro) processes as illustrated by Figure 3-3. Some macro processes found in businesses might be order fulfillment, billing, and product design. Micro processes would be activities like checking order specifications, preparing invoices, and drilling holes. "Micro" and "macro" are useful concepts, but they are only relative terms. Unfortunately, there is apparently no term for everything in between.

The reason for looking at measurement from a production process perspective is that making a product or providing a service is accomplished by processes, which cut across departments or other operating units. As Figure 3-4 illustrates, when viewed from the process perspective, the operations and their performance measures will make sense, but when viewed from a department perspective they will be seen only as isolated activities. Looking at work from a departmental viewpoint is a little like taking a dozen picture puzzles, mixing all the pieces together, giving several people a handful of the assorted pieces, and asking them to describe the result.

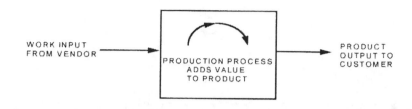

WORK INPUT
FROM VENDOR → PRODUCTION PROCESS ADDS VALUE TO PRODUCT → PRODUCT OUTPUT TO CUSTOMER

Figure 3-1   Every Business Activity is a Production Process

Undoubtedly, their reply would be: "This doesn't make any sense" and they would be absolutely right. Most processes don't make much sense when they are viewed from a departmental perspective. Indeed, it is easy to find front line employees, supervisors, and even managers, who have little idea where their work originates, who the next customer is in the process, and what that customer wants in terms of quality of the product. Some don't even know what they contribute to the product or even what the final product looks like. In one case, which I personally witnessed, a machine operator made parts for forty years, but didn't find out where they were used until he was leaving for retirement.

This poor understanding of processes is the reason why a few leading companies have designated some senior managers to be "process managers." The primary duty of these managers is to understand how a key business process works and optimize its performance by breaking down departmental barriers. The objective is to do what is best for the process as a whole, not what is best for each department. Designated process managers are few in number today, but their numbers will probably increase as the value of this approach becomes more apparent.

If every business can be broken down into a set of processes, then measuring performance in a company boils down to determining how to

Activity	Vendor	Value added	Customer
Taking telephone order	Customer	Order in company database	Company
Answering inquiry	Customer	Question answered	Customer
Drilling hole in panel	Panel supplier	Hole in panel	Next step in process
Designing chair	Marketing	New product design	Manufacturing
Making sales call	Company	Information about prospect	Company

Figure 3-2   Activities as Production Processes

MASTER PROCESS

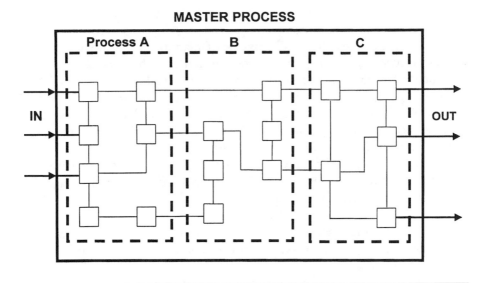

Figure 3-3   Any Business is Made Up of Small and Large Processes with Multiple Customers and Vendors

measure a production process. Since the output of a production process is a function of both the input and the process itself, the performance of a process cannot be determined by only measuring its outputs; its inputs and its internal operations must also be measured. Not measuring or being aware of the quality of the inputs to a process is probably the biggest

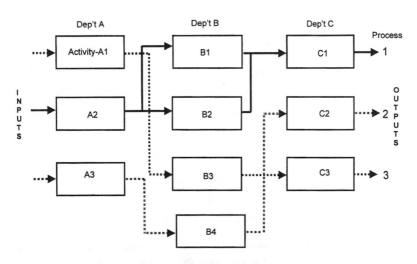

Figure 3-4   Work Gets Done By Processes Which Cut Across Departments

mistake managers make when evaluating the performance of processes, departments, and individuals.

To correctly assess the performance of manufacturing, for instance, it is not sufficient to know only its performance with respect to quality, productivity, delivery, and waste. These performance factors might be relatively poor, but does that mean manufacturing is performing poorly? Not if vendors are providing poor materials and late deliveries, the engineering drawings are late and full of errors, and sales keeps changing design specifications and required delivery dates.

Figure 3-5 shows the inputs and outputs that must be measured to understand *how* a process is performing and *why* it is performing that way. Each variable is defined as follows:

> **Work input quantity** — The amount of work to be accomplished, such as invoices to issue, service calls to be made, or beds to be assembled.
>
> **Work (product) input quality** — The quality of the work units coming into the process. Are the sales receipts legible and complete, the service calls well-defined, and the parts for the beds to specifications? Product quality is defined as a quality attribute that belongs to something directly attached to, or a part of, the product.

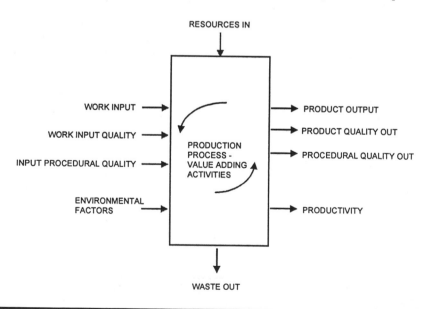

Figure 3-5   Measuring a Production Process

In most measurement applications, "quality" means how well a product or service meets its specifications. However, there are other dimensions of quality that are important to customers. The following quality attributes have been identified for products and services.

PRODUCTS[6]

- **Conformance** — how well the product meets its specifications.
- **Performance** — how well the product does what it is supposed to do.
- **Features** — how many options, bells, and whistles a product has.
- **Reliability** — the mean time between failure.
- **Durability** — how long the product lasts.
- **Serviceability** — how easily a product can be repaired.
- **Aesthetics** — looks, feel, and "sex appeal."
- **Perception** — how people feel about the product, as opposed to its real qualities.

SERVICES[7]

- **Tangibles** — what the customer sees in people, facilities, and equipment.
- **Reliability** — being able to perform the service dependably and consistently.
- **Responsiveness** — promptness and willingness to help customers.
- **Assurance** — making customers feel they can have confidence in the company.
- **Empathy** — conveying a caring attitude to customers.

One quality factor I would add to both lists is consistency. To some extent this is covered by conformance, but with some products, there is a broad range of what is acceptable if the quality factor is consistent. It is also true that if something is not conforming to a specification, it is generally easier to deal with the problem if the product is consistently wrong than if its quality is varying all over the place.

Some other types or categories of quality are given below. There are many different kinds of quality and it is important to distinguish between them. However, unless otherwise stated, "quality" means conformance to requirements or specifications.

- **Design quality** — What is designed into the product, i.e., plastic versus leather.
- **Execution quality** — How well the product conforms to the design.
- **Internal or process quality** — The efficiency and effectiveness of the production process.
- **External or outcome quality** — What the customer sees.
- **Strategic quality** — Doing the right things.
- **Operational quality** — Doing things well.

Input procedural quality — In addition to product quality problems, there may be problems that are not directly attached to the product. For example, sales receipts that are absolutely perfect except they are delivered too late to be included in the weekly computer run. Problems of this type are called procedural quality problems. Procedural quality problems are just as important as product quality problems in terms of productivity, costs, and customer satisfaction.

Product output quantity — The amount of product produced.

Product output quality — The quality of goods or services relative to the customer's requirements. Are the invoices correct, the service calls completed on time, and the beds properly assembled?

Output procedural quality — Procedural quality problems passed on to subsequent customers.

Resources consumed — Doing work must consume resources such as materials, labor, and energy. Most companies have good records of resources consumed. What they don't know is how many of those resources were wasted doing things that were unnecessary or correcting problems that never should have occurred in the first place.

Waste — Waste is any resource consumed that does not add value to the product. No process is 100% efficient, so it must generate some waste. Scrap and rework are common forms of waste in both services and manufacturing. Idle time of equipment and people, excess motion, work-in-process inventory, excess inventory, excess capacity, over-production, and unnecessary movement of materials or people are also forms of waste.

Most companies are wallowing in waste, but don't realize it, because the waste doesn't smell or attract flies. Another reason waste is not seen is that it is easy to equate being busy with productivity. Doing something that shouldn't have to be done in the first place is totally unproductive.

Productivity — Productivity isn't a tangible output of a process, but it must be measured to provide a complete picture of performance. It is defined as the ratio of work produced to resources consumed or output to input. Sales per employee and tons of steel produced per labor hour are measures of productivity.

Productivity is obviously very important to any business. However, low productivity is only a symptom of poor process quality. Consequently, quality is what should be emphasized when trying to improve performance, not productivity. Productivity is used as a cross-check on total process performance that will be described later.

"Measures of productivity do not lead to improvement in productivity."
W. Edwards Deming

Measuring productivity is a complex subject by itself, especially when trying to develop measures which combine labor and capital components.[8] However, because productivity is the result of high process quality, all that is needed for improving performance is an index that will reliably indicate which direction productivity is moving. Simple measures such as units produced per hour, shipments per labor dollar, dollars of billings per employee day, and equipment utilization will usually suffice.[9]

There are many ways to look at productivity and more than one measure of productivity may be required to provide a complete picture of what is happening from a total company perspective. It would be nice if only one measure was needed, but the laws of complexity and control make that impossible.

Environmental Factors — In most situations, there can be events or variables that are outside of the process but can still affect its performance. For example, the unemployment rate can impact absenteeism, turnover, and the length of time it takes to hire new employees. While it may not be possible to change environmental factors, they must be measured to understand why a process is behaving as it is. Knowing how external factors are changing is important for interpreting performance measures and may enable managers to compensate for the environmental changes.

Note that the value-adding activities are themselves production processes, which may need to be measured to understand how a given process works. In that case, the lower level measures can be thought of as diagnostic or process control measures for the outputs or outcomes of the larger processes.

The process model is simple, but important. It says that measuring these variables will describe how effectively and how efficiently any process is performing. The validity of the model is demonstrated in the following ways:

1. Any variable in any production process can logically be assigned to one of the measurement categories without having to make exceptions. If a production process is ever identified which has measures that cannot be logically put in one of the given categories, then the model will have to be changed.
2. All of the variables in the model can be found in any business process.
3. The model has been validated by using it to develop measurement systems that have been used to monitor and improve performance of business processes.

Others have proposed different measurement models,[10] but while they have some conceptual value, they do not reflect the way work is accomplished. They also do not provide a one-to-one relationship between the model variables and those that can be identified in the real world.

If all the variables shown by Figure 3-5 are measured, then a complete picture of process performance will be presented. If any elements of the model are missing, variations in the performance of the process will not be explained. All of the variables are important, but for improving performance, the quality variables are the most important. It is the product and procedural quality problems that cause low productivity, high costs, and dissatisfied customers.

Another reason why the quality measures are so important is that the average service or manufacturing company is spending 20 to 30% of its *sales dollar* on quality costs (appraisal, prevention, internal and external failure).[11] This is a huge pile of money in any company. It is about the same size as the gross margin in many companies and actually greater for some. These costs will never be reduced to zero, but even a 20% reduction will do wonders for the bottom line.

Quality variables (including waste) are generally the most important, but that does not mean the other variables should be dismissed as being unimportant. Every company and process is somewhat unique, so what is important in one situation may not be in another. Determining what is most important to a company, its customers, and its key production processes is the crux of the measurement problem.

MEASURING SERVICES VERSUS MANUFACTURING

It is often said that it is easy to measure manufacturing performance, but services cannot be measured. However, if every business activity is a production process and the performance of production processes can be measured, it follows that services can also be measured. Granted, there are some differences between manufacturing and services but there are many similarities as well.

One big difference between manufacturing and services is that goods have properties and functions they are supposed to perform that can be expressed as physical measures. Services and their quality characteristics are not often associated with physical measures other than delivery and response times. However, producing goods and services have many more similarities than they have differences. These similarities are

1. In both cases, the product is the result of a series of steps that are measurable production processes. The quality of these steps is also a leading indicator of customer satisfaction. Even with relatively intangible services such as interior decorating, legal services, or public relations, ordinary things like meeting schedules, answering the phone, returning calls, providing correct invoices, and doing what you say you will do, are very important to customers.
2. Both have "soft" quality properties that can be quantified, although it may not be easy to accomplish.
    *   Goods — features, aesthetics, perception, and satisfaction
    *   Services — assurance, empathy, and satisfaction
3. Both have "hard" quality properties, which can be relatively easily measured.
    *   Goods — performance, reliability, conformance, consistency, durability, serviceability
    *   Services — reliability, responsiveness, and tangibles (at least to a large degree)

4. Some services, such as custodial and repair services, deliver a product that has physical properties. Similarly, manufacturing companies typically supply numerous services in addition to their goods. Technical support, repair, supplying spare parts, and product customization are a few common examples.

5. Expectations play a large role in customer satisfaction for both goods and services. If a car meets all its engineering specifications but doesn't live up to what the customer has been told to expect, the customer is still going to be disappointed.

6. Customers' wants and needs for both products and services can be difficult to determine.

7. Customer satisfaction with goods and services can be measured by surveys and by indicators such as repeat order rate and customer retention rate.

Since a service must have customers, their satisfaction (an outcome) can be determined. Then if the service product factors that are important to customers are identified, the processes that create those characteristics can be determined. Once the key processes have been identified, measuring service performance becomes a matter of measuring the processes.

For manufacturing companies, direct labor costs have been steadily declining as a percentage of value added for the past century. Since 1945, direct labor has gone from 40% to less than 25% of value added over the production-merchandising chain. Across the spectrum of U.S. industry, manufacturing overhead averages 35% of production costs; the comparable figure for Japanese products is 26%.[12] If manufacturing companies need to pay more attention to overhead costs, certainly service companies need to do the same. The best way to reduce overhead costs is to keep them from growing out of control in the first place. Measuring the performance of overhead and administrative functions is the key to keeping these costs under control.

Quality and productivity of overhead and administrative functions can be measured. Certainly some creative services are abstract and perhaps very difficult to measure, especially when they produce one-of-a-kind items. However, any service being provided on a recurring basis is quite another matter. It may not be possible to measure some of the qualities of services as precisely as physical dimensions of a manufactured product, but they can be measured well enough to indicate what matters most — changes in performance.

Applying the Measurement Model — An Example

To see how the process measurement model can be applied, consider a human resources department of a discount department store chain. One of its functions is to acquire and screen candidates for store manager and department manager positions. The need for candidates to fill the different positions comes from two sources: new stores that are to be opened and turnover in existing stores.

The requirements for positions to be filled come from regional vice presidents, who are supposed to specify any special skills and experience required for the positions that are not contained in the standard position specifications. Upon receiving the position requisitions, the department reviews the applications it has on file. If additional applicants are needed, advertisements are placed in local newspapers. The candidates that pass the initial application review are personally interviewed by a specialist from the department in the store's respective city. Those that pass the personal interview and background checks are then presented to the regional vice presidents or store managers for final selection. The human resources department is supposed to have at least twelve weeks lead time to fill store manager positions and eight weeks for department managers.

Applying the measurement model to this process, the variables given in Figure 3-6 are the primary items that should be measured to provide a complete picture of how the candidate supply process is performing. Some of the measures given are stated in general terms, but more specific measures could be given if the details of the processes were known.

Note that these variables only address the inputs and outputs of the candidate screening process. The measures needed to monitor the internal workings of the screening unit must come from analyzing each of the sub-processes. Some of the process control variables that might be appropriate to measure are paperwork quality as it flows through the process, time delays in completing processing steps, work backlogs at various stages, and success rates in contacting candidates. In a real situation, more detail would be known about what the customers wanted as well as the internal workings of the department, so the measures could be defined more precisely.

In any case, a manager having these measures would have a good sense of how the screening unit was performing and to what extent work input and environmental factors were affecting performance. If productivity was deteriorating, the manager could look at the incoming workload quantity and quality, the waste factors, and the environmental factors, to identify what they were contributing to the problem. If internal process measures were available, they would identify what portion of the problem was created within the department.

In this case, a monthly summary of performance would probably be sufficient. However, the internal process control measures such as work backlogs should be monitored every week because spurts in demand or processing problems could develop in a matter of days. A more thorough

Model parameter	Process variable
Work input	• Store manager positions plus department manager positions to be filled each week or month.
Input product quality	• Accuracy and completeness of position requisitions
Input procedure quality	• Timeliness of requisitions
Product output	• Number of candidates for each type of position

Output product quality	• Candidates' compliance with procedures — have they gone through all required steps?
	• Conformance of candidates to specification
	• Conformance of candidates' records to specification
	• Quality of candidates could be indicated by:
	    Acceptance rate = number selected/number interviewed by regional VP
	    Turnover, the percentage that stay over one year

Output procedure quality	• Sufficient quantity of candidates delivered • On-time delivery of candidates • On-time delivery of candidates' records
Resources	• Labor hours, labor costs • Travel costs • Testing, advertising, communications and similar costs
Waste	• Under-utilized time in department (missed appointments, low activity periods) • Repeating tests and background checks • Candidate dropouts
Productivity	• No. of candidates delivered/$1000 of unit's cost • No. of candidates offered position/$1000 of unit's cost • No. of candidates staying longer than X months/$1000 of unit's cost **Note:** A weighting factor would have to be applied for the type of positions to be filled, since acquiring store managers would require more resources than department managers.
Environmental factors	• Unemployment rate — national and in locality of store • Location — distance from headquarters to the store could have significant impact on travel costs and utilization of interviewers' time. • Turnover rate of managers

Figure 3-6   Performance Measures for the Candidate Screening Process

knowledge of how the process works is required to determine lower level process control measures.

LOGICAL RELATIONSHIPS AMONG MEASURES

Logical relationships exist between performance measures that must be understood to correctly interpret and use them effectively. As has already been discussed, work is accomplished through processes and the production process is the basic building block of performance measurement. For improving quality and productivity, the process perspective is the most applicable, but there are other perspectives that are important in managing a business.

The Department Perspective

Figure 3-7 illustrates a production process where several departments perform different tasks in sub-processes. Although the process measures will describe how the process is working, companies are organized by departments or disciplines. Consequently, performance must be reported on a departmental or operating unit basis in order to establish account-ability and provide relevant feedback to work groups and individuals.

Since a department may have several customers, its performance must be measured by how effectively and efficiently it is serving each of them. The performance of department A is the "sum" of its performance in tasks A1 to A4.

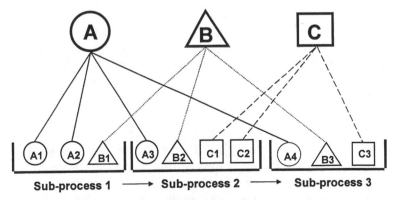

The performance of each department (A, B, C) is determined by how well it performs each task in different sub-processes for different customers.

Figure 3-7   Measuring Department Performance

For example, a quality index (Figure 3-8) could be constructed for department A by measuring the quality of each of the department's outputs and assigning weights to the different outputs based on their perceived value to the company. In the case of department A, the composite score could be developed as follows:

TASK	QUALITY INDEX (%)	WEIGHT	VALUE
A1	90	2	1.8
A2	70	10	7.0
A3	100	4	4.0
A4	95	7	6.7
Total	84.8	23	19.5

Department quality index = 19.5 / 23 = 84.8%

Figure 3-8   Calculating a Quality Index for Department A

This would require making a judgment about the relative importance of each task. Someone might argue that assigning weights is purely arbitrary, making an aggregate measure for a department constructed this way invalid. That could be true if the internal and external customers' requirements and the sub-processes were not understood, but in that case, none of the measures would have any meaning. In practice, determining the relative importance of outputs and quality characteristics may require some analysis and deliberation, but it is not that difficult to arrive at a useful consensus. In any case, a ballpark approximation is all that's needed because:

1. The trend and change in the performance measure is what's important, not the absolute value. If a weighting factor is off twenty-five percent, it is not going to make that much difference in observed change or longer term trends.
2. Managers should review the quality measures for each task along with the composite index, so any large deviations in the task measures will be seen and not hidden in the composite measure.
3. As the measures are used, better insight into the process will be gained and adjustments in weights can be made.

While it is possible to construct aggregate performance measures for a department, it is not a common practice, even in companies with good measurement systems. Instead, the key performance factors are monitored separately and managers rely on their knowledge of the relative importance of each measure to assess overall performance. Composite measures are discussed in more detail in Chapter 6.

The Customer Perspective

Customers generally don't differentiate between products or divisions when forming an opinion about a company. It is all one company to them, and poor performance in any one area can ruin the reputation of the whole company. Consequently, if managers want to know how their company looks to its customers, it must look at everything the customer sees — and that applies to performance measures as well.

An example is a medical and pharmaceutical products company that provides hospitals with different products and services from several divisions. From the hospital's perspective, it is all one company and there were incidents where poor performance from one division made it difficult for other divisions to sell their products to a particular hospital.

To get a clearer picture of how the company looked to its customers, the company developed a measurement and information system that monitors such factors as delivery performance, complaints and how successfully they were resolved, billing accuracy, response to inquiries, and the quality of sales support. This data is collected from all divisions serving a particular customer. Feedback about services and products is also solicited during contacts with customers and entered into the database, where it is converted into numerical scales.

With this database, which also contains the customer's purchasing history, managers can get an overall view of how well any customer is being served. Comparisons can be made between the company's different operating units and between customers as well. This performance measurement system has enabled the company to identify service and product problems along with opportunities to increase sales through better service. Standards have also been established for the performance measures that are most important to customer satisfaction.

The company's ability to see through the customers' eyes has improved its internal performance, customer satisfaction, and profits. In many cases, additional sales can be directly tied to improvements in product and service quality. In this case, the measurement system is complicated because of the many sources of data, but the calculations involved are quite straightforward. It took some time to develop techniques that would consistently quantify the subjective data and to determine the relative importance of customers' requirements, but the system was providing some very useful information in less than six months. The investment in developing the system and the costs of operating it are vastly outweighed by the benefits.

Most companies have some information about how their customers view them through returns, allowances, and complaints, but this is only the tip of the iceberg. As a salesman in a building materials company once told me:

> *"We have all kinds of problems around here. We get complaints about delivery, product quality, sales support for making bids, and billing all the time, but the information never gets to the CEO. If he knew what was going on around here, he'd have a fit."*

He probably would. But why is he not seeing what the customer sees? The reason may be that the managers who report to him don't feel the problems are important. The managers may also be out of touch with their customers. Another possibility is that they could be deliberately covering up the problems to avoid unpleasant repercussions from the CEO.

However, the most likely reason why the problems don't get communicated to the right people is that there is no organized system for collecting and summarizing all the feedback from customers. Without a structured system in place for collecting the data, it is unrealistic to expect anyone to know how well customers are being served.

Serving customers is what business is all about. When customers don't get what they want, they go someplace else. Furthermore, they don't come back unless their new suppliers fail to satisfy them. Knowing how a company looks to its customers can save the high costs of having to replace them.

The Company Perspective

As stated before, a company is a complex set of processes with multiple inputs, production processes, outputs, and customers. These inputs, outputs and customers are shown in Figure 3-9. As the figure shows, a company has four distinct groups of customers, each of which might have several sub-categories. For example, for retail products, customers could include the retail customer, the distributor, the store chain, and the individual store manager — all of which might have some special product or associated service requirements to be satisfied.

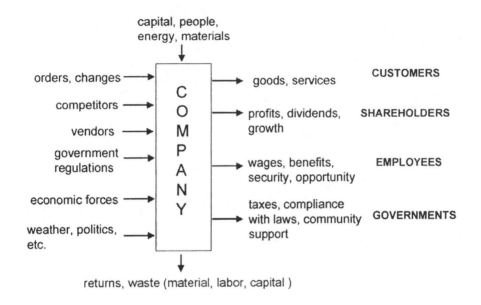

Figure 3-9   The Company as a Production Process

To measure the performance of this model company, we have to determine how well it is serving its different customer groups. Examining the customers' requirements, we might arrive at the list of key performance measures given in Figure 3-10.

Looking at the company as a production process brings home the point that what is important to the company (the shareholders) must also be included in determining what to measure. In a sense, a company is its own customer with its own interests. A good performance measurement system must reflect what is important to all of its customers: the company, its customers, its employees, and arguably, governments and society as a whole.

Another vital point is that although financial measures are critical to a company's success, they only address a portion of what is needed to understand how a company is performing. At best, accounting measures can provide some insight into operational performance, but they rarely go much further than indicating a problem or opportunity exists.

A company is very complex and many measures similar to those given above are needed to understand how it is performing. Success comes from both having a good strategy and executing it well, but how can you determine if a strategy is correct if a company's operational performance

CUSTOMER & INTERESTS	PERFORMANCE MEASURES
Consumer Quality, value	• Customer satisfaction index • Company ranking relative to competitors • Repeat purchase rate, dropout rate • Complaints • Product and service quality • Returns and allowances • Service call rate
Shareholder Profits and dividends Growth	• Financial strength (see accountants for details) • Profit margin • Labor productivity • Equipment utilization • Inventory turnover • Sales/employee • Sales growth rate • Order backlog (a short term measure) • Market share and rate of change • New product introduction and success rates • Percent of sales from products less than X years old • Weighted average age of products sold • New product development rate • Length of new product introduction cycle • Organizational strength - knowledge, skills, abilities • Capabilities in chosen areas of excellence
Employees Wages, security, opportunity, fair treatment, recognition, reward	• Wage rate ranking in local area (including benefits) • Turnover and absenteeism rates • Promotion rates • Satisfaction survey index • Training investment and participation • Knowledge, skills, and abilities
Governments	• Taxes • Jobs created • Payroll growth rate • Capital invested in plant, equipment, training • Violations of laws and regulations

Figure 3-10   Key Performance Measures for a Company

is poor or unknown? Unless a company's key production processes are under control and performing well, implementing a new strategy may be the wrong thing to do. Even if it was the right step, weakness in operations could make a new strategy unworkable. For that reason, operational performance should never be an unknown quantity.

Before you decide where to go,
first determine where you are.

The Strategic Perspective

Developing a strategy requires a company to choose which customers to pursue and how it is going to compete for them. It is the answer to the question: "What do we have to excel at doing in order to survive and thrive?" Every company has a strategy whether it is documented or not. Assuming a company's strategy is more specific than doing whatever feels good at the moment, it should have some strategic objectives.

Those objectives need to be translated into what is most important for a business unit to accomplish and be reflected in the unit's performance measures, but that is not enough. Everyone in the unit must understand how their performance measures are linked to the company's strategic and operational objectives, and their relative importance. If done properly, this alignment of performance measures with strategy will have everyone working toward the same objectives, eliminating costly conflict, duplication, and voids.

It should be possible to start with a strategic objective and trace a path showing how it is supported by all appropriate units in a company. This also means that some departments and other operating units should have some performance measures that are directed at strategic goals other than operating performance. As the Japanese auto manufacturers have demonstrated, operational excellence can be a powerful strategic weapon by itself, but there can be strategic objectives that are not directly related to operational performance.

As Kaplan and Norton point out in *The Balanced Scorecard*, identifying customer needs, creating the product, building and delivering the product, and providing post-sale service are all part of the chain to provide value for customers.[13] Since operations is only part of the chain, improving existing processes is not sufficient to assure a company's success. A company must also decide what market segments to serve, determine what its customers want, and respond to those requirements.

In developing its strategy, a company may find improving existing processes addresses many requirements of its customers. However, it may also discover that some processes are not needed and some customer requirements require new processes. For example, one company discovered its most important customers wanted rush deliveries that its normal delivery methods could not provide. What was needed was a new process, not just improving the existing one.

Segmenting markets, determining customer requirements, and measuring customer satisfaction as well as the processes that provide the deliverables have been part of the Baldrige award criteria since its inception in 1988. Although improving existing production processes is a key part of the Total Quality Management concept, it has never been the only requirement. "Quality starts with the customer," is the fundamental principle of TQM, with strategic quality planning and leadership probably being the next most important factors. Some managers may feel that quality improvement is only concerned with addressing current processes, but they are mistaken.

All process measures are linked to each other in some fashion because all parts of a production process contribute to the final product. The measures are interconnected and follow a logical "tree" relationship as shown by Figure 3-11. Drawing a cause-effect tree for a problem or outcome is fundamental to understanding how a process works. It is one

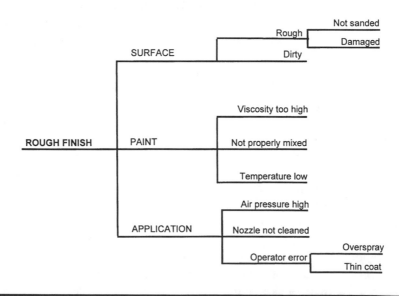

Figure 3-11   Cause-Effect Relationship Tree

of the techniques for determining what should be measured that will be discussed in Chapter 5.

One observation that can be made from the cause-effect tree is that no performance measure truly stands alone. Any single measure must affect a higher level measure or be affected by a lower level measure. This is certainly not a startling revelation, but it highlights the point that the different functions within a company are inter-dependent not independent. Some of the relationships between units of a company are very weak and some are very strong, but you would never know it from the way many companies still operate. Rigid "silos" are constructed between units and everyone operates as if they are in their own little world. Once any work is passed over the wall, it becomes the other units' responsibility, even if there is something wrong with it. Performance measures will make the dependency relationships between departments much more apparent. This should at least lead to some reduction in the height of departmental silos and perhaps their eventual destruction.

Both the process perspective and the cause-effect perspective say performance measurement must be accomplished by an integrated system of measures. Performance measures cannot be chosen on a haphazard or piecemeal basis. They must logically fit together and reflect the steps and cause-effect relationships of a production process.

The Perspective From the Nth Dimension

No, this isn't about the view from a UFO, but how performance looks from the many "dimensions" or variables that can affect a process. For example, the quality of units in a furniture manufacturing plant may vary with the model number, operator, machine, supervisor, production line, time of day, or any number of other variables. Looking at the plant as a whole, will show one set of problems; looking at each department within the plant, will show another. So too will summarizing performance by each process variable such as model number, machine operator, or type of quality problem.

Summarizing performance data by each of the different variables that could affect a process is called "stratification." This can reveal important relationships between variables and help identify the causes of problems. If we looked at product defects in an appliance plant as a percentage of units produced by model, shift, or production line, we might see this breakdown (Figure 3-12).

Model	Percent	Shift	Percent	Line	Percent
AX123	18.7	A	38.8	1	23.3
BZ777	38.4	B	30.7	2	24.8
AX238	23.3	C	30.5	3	26.7
CD765	19.6			4	25.2

Figure 3-12 Defects by Model Number, Shift, and Line

Although this data may not be conclusive, it indicates particular problems may be associated with model BZ777 and shift A. This doesn't say there is a cause-effect relationship or identify the cause of the problem, but it does say more investigation is in order.

SUMMARY

Every business activity is a production process that takes work from a vendor, adds value to it, and passes the product to a customer. Since every business is a complex set of production processes, it follows that for the purpose of improving performance, determining how to measure a production process is the crux of the measurement problem. That question is answered by the measurement model, which can be applied to any production process, whether it is producing goods or services.

Although the process perspective is the most meaningful for measurement purposes, there are other perspectives that are equally important. The department perspective is essential for establishing accountability. The customer perspective enables a company to see itself through its customers' eyes. The company perspective identifies what is important to the company and the strategic perspective shows how company objectives are supported by all of its units.

The cause-effect perspective shows that all measures are linked together. Consequently, performance measurement must be accomplished by an integrated system of measures that reflect the entire process, not just a few measures chosen because of their financial impact.

The different ways in which performance measures and their related data can be viewed provide different information and insights that are relevant for different decisions. However, all of the data comes from the same source — the production processes of a company.

4

OPERATIONAL REQUIREMENTS FOR EFFECTIVE MEASUREMENT SYSTEMS

Measuring all of the appropriate process variables, is not all that is required to have a performance measurement system that will be effective in improving performance. There are additional technical and cultural requirements which must be satisfied for performance measures to provide managers with accurate, relevant, and timely information.

TECHNICAL REQUIREMENTS

Wholeness

"Wholeness" means that all of the variables needed to completely define "good performance" are measured. Wholeness is necessary to provide a complete and balanced picture of performance so that performance won't look better or worse than it actually is.

For example, one company established a goal for its manufacturing manager of reducing the percentage of production orders that were behind schedule to less than 5%. The actual performance measure used was the number of production orders that were more than five days behind schedule divided by the total number of released or active production orders. Assuming the schedules were reasonable to begin with, this seems like a good indicator of how well the plant is adhering to production schedules.

It took only two months for the manager to reduce late production orders from 11% to 8%. In two more months, late production orders were reduced to 6%. However, raw material inventory was climbing out of control for no apparent reason. At the end of four months, raw material inventory had increased 30%.

Analyzing the production and inventory data revealed that the total number of released production orders had more than doubled, even though sales had not increased at all. Production orders were being released long before they were needed, which, in turn, were triggering raw material purchases that were just sitting in the warehouse. It was also discovered that the number of late production orders had not changed at all, but the percentage late had decreased because there were more released orders in the system. In fact, if the released, but not active, production orders were not counted as being released, the percentage of late production orders was still at 11%!

What was happening, was that the manufacturing manager was releasing more production orders into the system, making the denominator in the late percentage calculation larger and the percentage of late production smaller. He was making himself look good at the cost of flooding the warehouses with inventory and destroying the integrity of the production planning system.

This costly situation should never have happened in the first place. It takes a special person to manipulate figures like that, but similar events are not terribly unusual when there is intense pressure to reach objectives. In any case, the problem could have been avoided if the measurement system also monitored variables such as the number of released production orders (work input), the ratio of active to inactive released orders, or the number of orders that had been released earlier than necessary (those that had a delivery date greater than X weeks later than the current date).

As this case illustrates, performance measures can drive behavior in bad as well as good directions. Managers must be careful in establishing performance measures and goals. If they aren't, they will invariably get what they asked for, but didn't really want. The only way to avoid this problem is to carefully consider both the objectives and the constraints under which objectives are to be achieved. An objective should always be stated as: "Maximize or minimize a performance factor subject to condition A, B, C, etc." The measurement system must then include the variables that will present the complete picture of performance, not just a few pieces of it. In essence, the performance measures should define a multi-dimensional "box" where the process must operate while a given variable is optimized.

Understanding how a process works and how a performance measure is calculated is also beneficial when setting objectives. In this example, if the person setting the objective took the time to understand a few fundamentals about the production planning and control system, he probably would have changed the way the goal was defined and measured.

Objectives that are based on one performance measure are inherently risky. If all the elements of the production process measurement model are in place (see Figure 3-5), that is generally sufficient to provide a complete and balanced picture of performance. However, a process can have many inputs and outputs and some can be overlooked. Sometimes, despite valiant efforts, the missing pieces have to be found the hard way, but that is just part of the game of developing a good measurement system and establishing objectives.

Explain the Performance Gap

Knowing there is a difference between desired and actual performance is one thing, but in order to reduce the performance gap, its causes must be understood. To accomplish this, a measurement system should be able to break the performance gap down into components that would explain at least 80 to 90% of the difference. The 80 to 90% factor is derived from observing that generally something in the order of 20% of the causes of a problem, account for 80% or more of the deviation in performance. The other 80% of the causes are minor contributors that could be called "miscellaneous." If 20%, or less, of a performance gap is unexplained, it is unlikely that any single item not identified and measured is very significant. In practice, explaining 95% of any performance gap in specific terms is normally not difficult to accomplish.

In Figure 4-1, the 100 hours of equipment downtime is explained by the three larger categories of problems, which are broken down into eight more specific causes. These eight causes account for 95% of the downtime, leaving 5% as miscellaneous or unexplained. With only 5% unexplained, we can be confident the primary causes of downtime have been identified.

Sufficient Detail

Getting sufficient detail about quality problems is important for three reasons:

1. **To isolate the cause of a problem to the point where action can be taken.** General problems cannot be solved because they

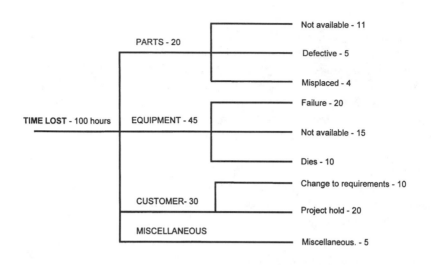

Figure 4-1   Explaining the Performance Gap

are only a symptom of lower level problems. Referring to Figure 4-1, if all that is known is 100 hours of time was lost, about all that can be done about it is complain or encourage everyone to do a better job — which will probably have no effect at all.

Breaking the 100 hours down to the eight problems gets closer to where action can be taken, but even here, problems such as "failure" and "dies" provide little useful information. What failed? What was wrong with the dies? Without more detail, general solutions like buying more dies or increasing maintenance on the equipment can be applied, but they may not address the real problem. At least one more level of detail is required to get to the actionable level and two or more additional levels of detail might be necessary.

Only specific problems can be solved. A general problem such as defective parts should ideally be broken down into the individual types of defects and perhaps even the different causes for each type of defect. It may not be practical to get all the way down to the root cause of every problem, but getting to a specific step in the process is normally a minimum requirement and is not difficult to accomplish.

To demonstrate the need for detail, consider the problem shown by Figure 4-2. Spend about three minutes on it and then see the box on page 50 for the answer.

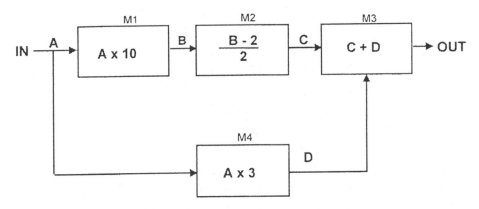

PROBLEM: With an input of (1), the output should be (7), but the actual output is (5). What modules are defective?

Figure 4-2 Detail Problem

Try solving the problem again, but this time you are given the additional information that value at point B is 10, at point C it is 3, and at point D it is 3. See box on page 54 for the answer.

2. **To be able to establish accountability for a problem.** As will be discussed later, accountability must be established to determine who should be responsible for improving a performance measure or solving a particular problem. If enough detail is provided to identify the cause of the problem, it will be possible to establish accountability. However, it is possible to establish accountability to a department, work group, or individual level without being able to identify the cause of the problem.

3. **To be able to see all problems and opportunities for adequate control.** As Figure 4-3 illustrates, large variations in lower level measures can be completely masked at upper levels. For example, overall on-time delivery performance might look good, but a closer look could reveal problems with particular products, customers, or sales representatives. Average figures can hide numerous problems and opportunities. A man with one foot in ice water and the other in water at 140°F may, on the average, be comfortable, but he will actually be in considerable pain.

Complex processes can only be controlled by systems that have at least the same degree of complexity. For example, if a company has 1000

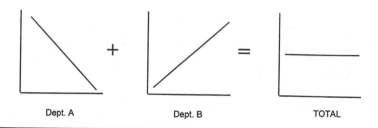

Figure 4-3 Insufficient Detail Hides Problems and Opportunities

different items in inventory, it cannot control inventory very well by lumping products together into five separate groups and ordering all items in the group at the same time. For proper control, each item must be treated separately and ordered at the appropriate time. (It is conceivable that 500 of the 1000 items could have the exact same demand and supply characteristics, but in that case, from a complexity viewpoint, there are only 501 items, not 1000.)

Businesses are very complex systems, so for proper control, their performance measures must reflect that complexity. Operations at the bottom cannot be controlled by measures at the top. This typically means getting to details several levels below the top level performance measures a company may have in place. Isolating problems to a department is usually not sufficient, because there are additional levels below the department such as the process step, the problem, and the different causes of the problem.

The need for detail is at odds with every manager's desire for quick and easy solutions to complex problems, but that is a fundamental mismatch that only works in management fairy tales. As Mark Twain put it:

"For every complicated problem, there are at least three simple solutions — all of which are wrong."

**Answer to the first part of the defective module problem —
Figure 4-2**

The answer to the problem is there is no answer. The problem cannot be solved with the available information. If you were faced with this problem in the real world, the only way you could solve the problem would be to substitute combinations of modules until the correct output was achieved. Then the solution could be verified by applying other inputs and checking the output.

This does not mean a measurement system's effectiveness is determined by how much detailed data is collected. To the contrary, collecting useless detail will just waste resources and perhaps cloud the issue of what really counts. How much detail is enough? It all depends on the complexity of the process. Where there are many steps in a process, many opportunities for error, or a high degree of variability in a process, more detail is required. Unfortunately, what appears to be a simple process can often be quite complicated, so it is not possible to provide any guidelines.

The best approach to determining the necessary level of detail is to follow the procedures given in the next chapter and initially get as close to the root cause of problems as possible, without grossly complicating the system. Then, using the system will determine if additional detail is necessary.

Accuracy

For improving quality and productivity, a high degree of accuracy is not required. What matters most is consistency in reporting, so any change in a performance measure is a reliable indicator of a change in performance. Reporting errors are bound to occur, but if the same people are reporting similar data day-after-day, the errors in the data will be quite consistent. This will make the error in the *difference* in performance at two points in time negligible. After everyone is trained in reporting and becomes familiar with the data reporting system, the most common error is failing to report an incident. This means that quality problems are normally understated a small amount, which is not going to affect any conclusions drawn from the data.

Of course, reasonable efforts should be made to make the data as complete and accurate as possible, but performance measurement systems don't have to pass financial audits. If 2% of the data is missing and what's reported is 97% accurate, that is good enough. Trying to achieve perfection, will only increase data collection and processing costs and delay producing reports.

> *"Far better an approximate answer to the right question, which is often vague, than an exact answer to the wrong question, which can always be made more precise."*
>
> **John Tukey**

Timeliness

What did you have for dinner yesterday? Certainly, you know.

Now can you tell me what you had for dinner two weeks ago? Unless you eat the same thing for dinner every day, you probably won't be able to answer the question without a great deal of thought — if at all. This illustrates why there should be as little time delay as possible between activities taking place and the resulting performance measures being available. A week is a long time for the human brain to remember any details that may be important for identifying the causes of problems.

Timeliness is important because understanding the circumstances behind the numbers is often essential in diagnosing problems. What does "timely" mean in business? That depends on the situation and how quickly things can change, but some general guidelines are

- In the average business situation, performance measures and related reports for a given week should be available the next day after the end of the week. If a workweek ends on Friday, the reports should be available Monday morning.
- Where there are a large number of transactions each day, such as in many service and manufacturing environments, summary figures should be available at the beginning of the next day. This, of course, does not apply to micro-process statistical controls, which can require very rapid response.

There may be some slow-moving situations where time delays of a week or two might be acceptable, but they are relatively rare. In the normally busy business environment, the ability of people to relate measures to previous events decreases significantly in a few days. Ancient history may be interesting, but if numbers can't be linked to the activities that created them, the information is not going to be as helpful in improving performance as it could be.

Frequency

A process must be measured with a frequency that is consistent with how fast it can change. As Figure 4-4 illustrates, if measurements are not taken often enough (frequency = 1X), the performance picture can be quite distorted, but if taken too frequently (frequency = 6X), the cost of measurement will increase without significantly improving the quality of the information.

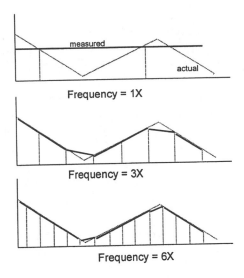

Figure 4-4   Measurement Frequency

A snail sliming his way through New York's Central Park was mugged by a pair of turtles. When the police arrived, they asked the snail to describe his assailants. He replied: "I'm sorry officers, but everything happened so fast, they were gone before I could get a good look at them."

As this story illustrates, speed is a relative matter. So, too, is the response time of a process and how often it needs to be measured. A good rule of thumb is that a variable should be measured a minimum of six times during the period in which it can experience significant change. Something that can change in a minute should be measured at least every ten seconds; what takes a week to change must only be measured every day. For some variables, quarterly or annual measurements through audits or other means could be sufficient.

The Learning Cycle

Another characteristic related to process response time is what I call the "learning cycle." This is how long it takes to make a measurement, implement a change to the process, have the process respond, and then obtain enough data to determine the effect of the process change. The length of the learning cycle varies with how long it takes to make particular changes, but its lower limit is defined by how long it takes to get enough data samples to reliably measure a change in performance.

Where customer satisfaction is concerned, there will almost certainly be a time delay between improving products and when customers will begin to change how they feel about them. Since reputations are not gained or lost in a day or two, it could take months or years for a large portion of customers to change the way they feel about a product. However, if customer wants have been properly identified, the process performance measures will be leading indicators of customer satisfaction. This could be verified by surveying customers who had received new products to get an early indication of the impact better quality will have on customer satisfaction.

The length of the learning cycle is important to consider when designing a measurement system. Where process response times and learning cycles are long, it is more important to measure lower level variables that will be leading indicators of process performance. The minimum learning period must also be considered when interpreting measures because it limits how quickly the effects of process changes will be seen in upper-level performance measures.

When someone says, "That can't be measured," what they are often saying is that changes made today in processes, procedures, or materials don't show up in tomorrow's results. This is often true, but it does not mean the effect of process changes can't be measured. It can take years to confirm the effectiveness of medicine or surgical techniques, so early judgments are made on the basis of samples from controlled experiments and clinical trials. Unfortunately, these measurement methods are not infallible, but the alternative is to seriously delay the introduction of better drugs and treatments.

Answer to the second part of the defective module problem — Figure 4-2

Knowing that the value at point B is 10 says that module M1 is working properly. Since the value at D is 3, M4 is also good. The value at C is 3, but it should be 4, so M2 is defective. If D is 3 and C is 3, the output should be 6, but it is 5. Therefore, M3 is also defective. An easy problem to solve — but only if you have performance data to an appropriate level of detail.

Systematic Operation

Business processes are like machines. They do the same or similar things day after day and because they do, they must be measured on a regular basis. This requires having a measurement system that is an integral part of the business, not special procedures used only when unusual circumstances arise.

"Systematic" means not only collecting data and measuring performance on a regular basis, it also means reviewing the information on a regular basis and using it to make changes in priorities, procedures, and how resources are allocated. This will be discussed in more detail in Chapters 7 and 8.

Long-Term Consistency

Recognizing short-term changes in measures is vital for improving performance, but it is also important to be able to make reasonably accurate comparisons over longer periods such as a few years. For this reason, the measures used must be insulated from changes in product mix, sales or production volumes, costs, and other factors that could distort the picture. Using ratios or percentages such as "errors/invoice line item," or "percentage of pages needing major rewrite" will eliminate the consistency problem and make comparisons between different periods valid.

Absolute units such as the number of errors, hours lost, or complaints received should never be used to measure performance, because these numbers will fluctuate with work volume. When that happens, the measures can be very misleading. A case in point is a company that promoted a decline in the number of quality discrepancy reports as showing an improvement in quality. In actuality, the only change the measure reflected was a decrease in sales.

One way to establish a consistent basis for determining performance measure ratios is to define a "standard" product and reference all work back to that standard. For example, a tire manufacturer could use a popular model as the standard and reference all the different types produced back to that same standard. Quality and productivity indexes could then be developed using the equivalent work units. Whatever method is used to determine the number of work units, it does not have to be perfectly accurate — a reasonable approximation that is consistently applied will work very well.

No matter how well planned, however, changes in product mix or processes may occur that can make using the established measures impractical. In that case, a new system must be used, but even then, it may still be possible to develop a conversion factor to link the new system with the old.

Financial Measures Versus Operational Requirements

When financial measurement systems are compared to the operational requirements, they generally fail to meet the criteria for wholeness, explaining the performance gap, detail, timeliness, frequency, and accountability. Newer accounting concepts such as activity based costing and value-added measurement provide additional insight into a company's operational performance, but even with these techniques, financial systems will not satisfy the operational requirements for effective measurement. They combine too many variables, speak the wrong language, and cannot economically meet the requirements for detail, timeliness, and frequency while also satisfying the rigid requirements for accuracy and traceability demanded by accounting standards.

This does not mean that financial measures are useless or somehow inferior. They are essential for managing the financial side of any business and provide some measures of operational performance, but they are not adequate for managing or measuring the operations of a business. Using financial measures to manage operations is like using a screwdriver to drive nails — it won't work very well because a screwdriver is the wrong tool. By the same token, operational measures can't be used to manage the financial side of a business.

Despite the obvious differences between financial and operational performance measures, many executives think their financial systems provide them with virtually everything they need to know. Unfortunately, they are wrong.

CULTURAL REQUIREMENTS

A performance measurement system may be technically perfect, but it will still fail to be effective if the social or cultural conditions within a company don't meet certain requirements. These requirements are not particularly difficult to satisfy, but they can be in conflict with a company's existing culture. If so, the company has a choice of adopting some new forms of behavior or forgetting about having a useful performance measurement system. The necessary cultural characteristics are as follows.

Absence of Fear

Absence of fear is one of Deming's Fourteen Points, but for performance measurement, there are two specific fears that will make a measurement system useless. The first is the personal fear of being reprimanded, embarrassed, or otherwise beaten by the measurement yardstick when the measures look bad for any reason. If there are negative consequences to being measured, there is every incentive to distort the data. Not many of us would volunteer to be whipped.

The second fear is the fear of getting one's co-workers or friends in trouble by reporting problems that reflect poorly on their performance. This is a close cousin of the first fear and may just be a way of covering up an individual's fear of doing damage to himself, but it is commonly expressed.

Unlike accounting systems where there are records to assure transactions are reported, most of the necessary data for operational measures will have to come from people voluntarily reporting the problems they encounter. If there is distrust of management and fear that the performance measures will be used as a "gotcha" tool, you can be sure many problems will not be reported or will be reported incorrectly. If the data has too many errors or omissions, the measures will not be credible and the integrity of the whole system will be destroyed.

How many errors are too many? Based on experience, if 5% of reported data is incorrect, some credibility problems will result; at 10%, anyone could cast some doubt on any measure, creating serious credibility problems. An omission rate of 5 to 10% could be tolerated if the omissions are distributed evenly throughout the process. In practice, however, concentrated pockets of omissions that could significantly bias the data, is the most prevalent problem. This can be corrected with training and supervision.

Accountability

Accountability hasn't received much attention in books about quality improvement or, for that matter, management in general. It is my experience, however, that accountability is an essential ingredient for producing quality work and improving performance. Of course, most people assume some responsibility for performance, but not necessarily all they should. This is especially true when a performance problem is the result of several factors controlled by different departments and no one is clearly accountable for the whole problem.

There is a big difference between saying, "*We* have a problem" and saying: "There is this performance problem and this part belongs to department A, this part to department B, this part to department F, and so on." In the case where the problem is a general symptom, it would be a rare group of managers who would not deflect criticism of their department's performance by pointing out all the problems caused by every other department. Even if all the managers wanted to do what they could, it would be difficult for them to know what to do if they didn't understand the impact their department was having on performance. Without reliable information to the contrary, each manager might very rationally assume his part of the problem was minor and there was little he could do to improve the situation.

In the second case, the general problem is converted into specific components, which are "owned" by individual managers and departments. This eliminates finger-pointing and buck-passing because it is clear what each department is contributing to the problem. In addition, significant pieces of the problem can't "fall through the cracks" because there are no cracks.

> *"It's fourth down and two yards to go for the touchdown we need to win the game. Now I want everybody to go out there and do something good!"*

Can you imagine a coach calling a play like that? Some fans of teams that are in last place may think the coach is behaving that way, but I'm sure it has never happened. However, when managers don't define what good performance means and who is accountable for specific aspects of performance, they are not calling the play as they should.

Some people may equate accountability with beating people over the head and think it is contrary to the Deming philosophy. However, Deming never said accountability was a dirty word and one distinguishing characteristic of the Japanese culture is a strong belief in personal accountability. In what other societies do you see corporate leaders resigning or crying and admitting failure in public?

In any case, what I mean by "accountability" is not "It's your fault and you will hang for it." What it means is that:

> *"This problem is coming from your area and you are responsible for solving it. I do not necessarily expect you to solve it all by yourself, but I do expect you to understand the problem, to know what's being done about it, and to be the driving force to eliminate it. If you need help, resources, or something from other departments, it is your responsibility to tell me and I will do what I can to remove the obstacle. However, make no mistake that it's your problem."*

Managers, supervisors, and front line employees do not object to being accountable for performance, providing they control the outcomes to a large degree and they have reasonable authority to change the process to improve its performance. Holding someone accountable for what they don't control or for decisions they don't make is just plain stupid, but it happens all the time. Some instances that I have personally witnessed are

- An inventory control manager being blamed for having too much inventory, which was the result of a bad sales forecast.
- A sales manager being criticized for not increasing sales when the products he had to sell were poor quality in many respects.
- A manufacturing manager getting blamed for late deliveries when the real problem was late release of engineering drawings to manufacturing.

These and similar instances do a great deal of harm because they destroy the morale of managers doing their job and encourage game-playing by those who should be held accountable. What may even be worse for the company, is that the real problems never get addressed, so they are sure to happen again. Everyone's performance should be measured and everyone should be accountable, instead of blaming the person who happens to be the most visible in a process.

For accountability to be clearly defined, each and every performance measure must belong to one, and only one, owner. There can be no measures that are orphans or have multiple parents. The accountability equation is

$$\text{If } R = 1, A = 100$$
$$\text{If } R > 1, A = 0$$

This means if one person is held responsible, then 100% accountability is established, but if more than one person is held responsible, then there is no accountability. "Split" or "joint" accountability is an oxymoron. When people talk about split accountability, what they are really saying is, "We haven't figured out what's causing the problem or who should be responsible for it, so we'll just throw it on the table and see who picks it up." When that happens, the odds are overwhelming that no one will pick it up. Even if some brave person volunteers, he will probably not succeed, because the others will not volunteer to own their parts of the problem.

A case that illustrates the fallacy of split accountability happened to me when I was managing a manufacturing operation. One day, Margie, the editing supervisor, came to me very upset because one of her machines had been down for three hours. When asked what was taking so long to get the machine repaired, she said, "The two maintenance guys can't agree what the problem is. Bob is working on the printer and Arnie is working on the editor (which feeds the printer) and they can't agree about which machine is at fault. Each one is blaming the other for the problem."

I went to Bob and Arnie, discussed the problem, and got absolutely nowhere. Frustrated, I took a quarter out of my pocket and told Bob to call it heads or tails. He called tails and won the toss, so I said to him: "It looks like you are the winner. I don't care where the problem is, but you are now responsible for both the editor and the printer. The problem needs to be fixed right away for us to meet the production schedule. Arnie is going to go do something else." In less than thirty minutes, both machines were back in operation. So much for the theory of split accountability.

Performance measures without accountability attached to them are just numbers. They may receive some passing attention, but they will result in little action. Similarly, without performance measurement, there can be little, if any, meaningful accountability in a business. Everyone in a company knows they are accountable. That is not the problem. The issue is what are they accountable for? Performance measures can answer that question in clear, specific terms.

Establishing accountability with a high degree of accuracy, is the critical difference between measures that will improve performance and those that will just keep score.

Validity

If everyone using performance measures does not accept them as reliable indicators, they will have all the impact one ice cube would have on cooling the Sahara Desert. If performance measures are not accepted or trusted by the users, there can be only two possible problems:

1. The measures are not understood.
2. There are some faults in the performance measurement system.

In either case, the problems must be eliminated through training or changing the measurement system.

Introducing performance measures to an organization is like getting everyone to speak a new language. It requires training so everyone understands what the measures, charts, and reports mean and don't mean. It also requires quickly answering any questions and correcting any problems within the measurement system before its credibility erodes.

Validity is not a birthright of a performance measurement system. It must be earned by people using the measures and finding that they help them do their job. Even with a perfect system, it could take several weeks for everyone to develop enough understanding of the measures to have confidence in them. For an initially less-than-perfect system, it might take several months.

How do you know when validity is established? One sign is when questions and criticisms fall to negligible levels, but the real proof is when managers start using the performance measures. More will be said about this in Chapter 6.

Easily Understood and Relevant to the User

If the measures are not easily understood, they will be misused, or more likely, not used at all. This will require training the users, but the most important factors in making the measures easily understood are

- **Employ terms that are familiar to the users.** It is easier for people to relate to quality and productivity in terms of something they see every day, instead of abstract ratios. Don't talk in terms of "percent deviation from standard cost adjusted for the wholesale price index." Use terms like "hours of rework/chair." If measures can't be stated in simple terms, it means nobody understands them in the first place.
- **Give people only what is relevant to them in a way that reflects accountability, relative importance and logical relationships.** If you own stock in three companies, what are the first stocks you look at every day — those you own or those your friend at work owns? Interest in performance measures follows the same pattern — people are interested in what they do or what affects them. Let everyone have access to the rest of the measures if they want to look at them, but don't confuse the issue of accountability by giving everyone every report and graph produced by the system.

Easy to Use

It is important to make reporting the necessary data as easy as possible for everyone involved. Procedures for reporting data are often given to front line people to use without considering their problems in using them. Then when they complain, they are told "That's just the way it is. Do it or else." Consequently, "or else" happens, and much of the data is either missing or incorrect. Making it easy to do right and difficult to do wrong is the key to high quality and productivity in any business activity, including collecting data.

The same is true about using the information. If reports and charts are difficult to interpret and use, they will only be used when absolutely necessary. In that regard, it must be recognized that since different managers make different decisions, they need different information. This basic principle is unrecognized in companies that bury managers in piles of reports that contain some useful information for everyone but fit no one's particular needs.

The reports and graphs of a measurement system should reflect accountability and process relationships. Most of the time, this will require separate reports and graphs for different managers and work groups. More details and examples are provided in the following chapters.

A well-designed measurement system should enable managers to identify the key issues in their areas of responsibility in less than ten minutes. Making a system easy to use takes some effort and investment, but it can make the difference between a system that produces results and one that just consumes resources. The burden for making performance measures easy to understand and use should be placed on the measurement system and its designers, not on the people who will have to supply the data and use the information.

SUMMARY

A performance measurement system must satisfy certain technical requirements in order for it to provide meaningful information and be effective in improving performance. Wholeness, explaining the performance gap, providing adequate detail, accuracy, timeliness, frequency, systematic operation, and having long-term consistency may seem like a formidable list of conditions to satisfy. However, these requirements are not all that difficult to meet with a moderate amount of effort.

The social requirements of absence of fear, accountability, validity, and being easy to understand and use, are also not formidable objectives. Some managers may have difficulty with not using performance measures as a reason for punishment, but this is a rare problem. Establishing accountability can raise some questions about who should be accountable for particular measures, but these questions will get answered as the measurement system is used and production processes are better understood.

5

DETERMINING
WHAT TO MEASURE

Determining what variables to measure and how to measure them is not an exact science, but there are some general principles and practical methods that can be used. An overview of the general approach for determining the quality variables a company should measure is shown by Figure 5-1.

The steps for determining the quality variables to measure in a company or other business unit are as follows.

1. Identify the customers of the process. For the company's customers, there could be different groups of customers (market segments) who receive different products or have different quality requirements for the same or similar products.
2. Identify the key deliverables supplied to the customers. This assumes that all customer requirements are being addressed.
3. Determine the customers' quality requirements for each product. These are the output, or Level 1, quality measures.
4. Identify the processes providing each of the products.
5. Determine how each principal business process works and identify the variables that affect the quality of its products. These are the Level 2, or process control, measures for the principal processes.
6. Determine the appropriate measures for controlling sub-processes, to as many lower levels as necessary to provide sufficient detail for establishing accountability and to get as close as seems practical to the root cause of quality problems.

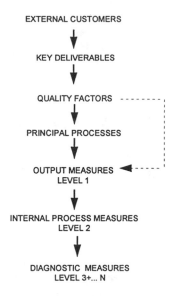

Figure 5-1   Sequence for Determining What to Measure

Determining What Customers Want

> *"The business process starts with the customer. In fact, if it is not started with the customer, it all too many times ends with the customer."*

William Scherkenbach

Whether customers are external or internal, there is no better way to determine what they want than to ask them. This can be accomplished in real-time, as customers are using a product, or on a historical basis after they have used a product.

Real-time feedback about likes, dislikes, and satisfaction can be obtained from customers by questionnaires, reply cards, problem logs, and on-the-spot interviews. Real-time feedback has the potential of providing thorough information about quality problems. It is limited by the actual problems encountered during a sample period, but this can be overcome by increasing the number of people involved or the duration of the survey.

If real-time collection of data from customers is not practical, after-the-fact interviews, focus groups, and surveys can be used to identify customer likes, dislikes, needs, and wants about both current and future products.

Since these methods depend on people remembering past experiences, it is possible some important issues may never be mentioned, even if the survey is properly designed. Detail about problems may also be lost when participants mentally lump similar events together, even though they are not truly identical. Regardless, after-the-fact surveys can still provide very useful information about customer requirements. In some cases, they may be the only option. Surveys and focus groups are the only way to get information about possible future products that customers have not yet experienced, although inferences can be made from feedback about other products.

This raises the question; "Do customers always know what they want?" When asked about a product or service they have experienced, the answer is "yes," at least most of the time. Where new products are concerned, the answer is clearly "not always." Although companies spend millions of dollars on market research, product development, and marketing campaigns, the success rate for new products is only about five percent. Of course, there are many reasons products fail in addition to companies not properly identifying what customers want.

In any case, there is no better alternative than asking customers and prospects to get information about their requirements. As Figure 5-2 shows, companies have many sources for acquiring that information and determining how well their current customers' requirements are being satisfied. However, it seems most companies don't make good use of the wealth of information about customer requirements they could easily obtain. One apparent reason for this is that much of the information that is obtained gets lost because there is no system for collecting and compiling the data. Another barrier is that there is often a strong incentive to resolve problems and complaints at lower levels and not report them for fear of making one's self or friend look bad.

When identifying customer requirements, two elements must be determined — the specific quality factor and its relative importance to the customer. Quality Function Deployment is a well-developed discipline for determining the relative importance of product attributes, but an extensive analysis of customer requirements is not necessarily required to start implementing performance measures. A good sense of what customers want can be gained from first compiling all the data available from inside the company (including interviewing anyone in direct contact with customers) and then having a third party conduct a satisfaction and competitive position survey.

A construction materials company that wanted to increase its market share by improving the quality of its products and associated services,

Information / Source	Satisfaction	Quality Requirements	Product quality	Value/price relationship	Company competitive rank	Customer plans, future needs
Customer service	■	■	■	■		
Salesforce	■	■	■	■		■
Management contacts	■	■	■	■		■
Company quality meas.			■			
Customer quality meas.		■				
Surveys	■	■	■	■	■	
Third party interviews	■	■	■	■	■	■

Figure 5-2 Common Sources of Information About Customer Requirements

provides an example of what can be learned from customers in a relatively short time. Although improving quality was a strategic objective, nobody really knew what customers wanted or what aspects of quality needed to be improved. Some customer requirements, such as on-time delivery, were readily apparent, but there were several critical areas where there were many different opinions about what customers wanted and very few facts.

In order to determine what customers wanted, everyone in the company having direct contact with customers (about 20 people) was interviewed to identify what they thought was important to customers. This information was then used to compile a list of topics and questions to use during interviews with customers. A list of typical customers, former customers, and prospects that had never been customers, was also assembled. Customer service personnel, who were familiar with the market, selected companies to be surveyed from the list to avoid any potential bias from sales representatives. When completed, the customer service team felt the list was a good cross-section of customers, former customers, and prospects.

One-on-one telephone interviews were then conducted with the individuals who made the buying decisions for the selected companies. This was usually the CEO or vice president of operations. These interviews were open-ended discussions that permitted exploring issues rather than just answering a fixed set of questions.[14]

After interviewing only twenty companies, it was possible to draw the following conclusions.

- There were two distinct market segments for the company's products. For one group, price was very important and quality considerations were secondary, providing minimum performance requirements were satisfied. The second group had high requirements for quality and service, but was willing to pay more to have these requirements satisfied, because this would lower their total cost of construction projects. The company was surprised by the findings about price sensitivity and the fact it was not meeting its most important customers' requirements for quality and service. Some non-customers mentioned this as the primary reason for using other suppliers.

- The quality conscious group felt that product consistency in terms of texture and color was very important to satisfying their customers. They also felt the company's performance in this regard was inadequate. This was quite a surprise to the sales reps. They believed that lighter color and weight were what the customers valued most in the product, but these characteristics were not at all important to the customers.

- An important service not being provided by the company, was the quick delivery of small quantities of product that could be required on short notice because of design changes, damage to materials, or other reasons. Because these minor shortages could cause major disruptions of construction schedules and significant additional costs, the customers were willing to pay a premium price for the product that would more than cover the cost of providing rapid response delivery.

- A major cause of dissatisfaction was the company's slow response in providing product specifications, price quotes, and technical support to customers when they were bidding on projects. This had nothing to do with the product or any service directly attached to it.

These findings were quite surprising to everyone. Not all companies would find they were so far out of touch with what their customers and potential customers wanted. In this case, knowing product consistency and rapid response to emergency orders was very important to the company's largest customers had strong implications for the company's strategy and what it should measure.

This case also illustrates that when soliciting feedback from customers and non-customers, causes of dissatisfaction as well as satisfaction must be identified. Satisfaction may not always make the sale, but dissatisfaction will almost certainly cause it to be lost if the customer has other vendors available.

One example of the impact dissatisfaction can have, was a public relations firm which received high praise for its work, but had little repeat business from customers. Interviews with current and former customers revealed that customers did not come back because the company had a poor record of returning phone calls, answering letters, and keeping customers informed of progress or problems with their projects.

With this knowledge, the company improved its internal communications system and implemented some simple measures and controls to improve communications with customers. After that, its customer retention rate was nearly 100%.

A capital equipment manufacturer experienced a similar problem, when it was advised that one of its major customers was about to sever relations. Management was shocked to hear this because they had never been informed of any problems with the equipment's performance. Some frantic phone calls revealed the problem was not with the equipment, but with the company's poor performance in predicting shipping dates and keeping the customer informed of any changes in shipping schedules.

The company reacted by implementing process measures that would identify potential production and scheduling problems. It also made changes to the production management system to provide a more accurate and timely view of work still to be completed on each order. These changes enabled the company to make reliable projections for shipping dates and to keep its customers informed. The company was soon back in good grace with its very important and almost lost customer. That was only one benefit, however. The feedback provided by the performance measures and the resulting understanding of the production process produced other process improvements that reduced order cycle time and costs. This was ultimately reflected in both increased sales and profits.

These examples illustrate two very important points. The first, is that companies probably have a poor understanding of what is important to their customers unless they have asked them. The second, is that operational excellence can have a powerful impact on customer satisfaction and sales.

Using an outsider is the best way to get reliable information from customers about their needs, wants, and dislikes for the following reasons:

1. Company personnel won't ask the hard questions they should ask and will tend to hear what they want to hear.
2. Company employees are not in a good position to get objective information from *non-customers*, which can be the most revealing information about where the company is failing to meet customer requirements. When approached by a company employee, former and prospective customers are likely to feel the company representative is trying to sell them something.
3. Few companies have any significant experience in conducting customer surveys. Like anything, it takes experience to know how to do it properly.

Key Performance Factors

Once the external customers' requirements have been identified, the next logical step is to define the company's key performance factors (KPFs). These are the performance factors that are most important from the customers' and the company's viewpoint. They describe where the company must achieve and maintain excellent performance in order to survive and thrive. While excellent performance is obviously desirable in all areas of a business, the KPFs define where it *must* be achieved, even at the expense of other performance factors.

In many respects, the difficulty in determining a company's key performance factors lies not in identifying things to measure, but in deciding what are the critical few items that will drive a company's strategy and its success. Is it more important to provide rapid response to orders or to keep inventory levels low? To personally answer calls within fifteen seconds or use answering machines with layers of menus to hold down costs? To offer low-priced goods and services or those high in design quality? Having high quality production processes can reduce the trade-offs that have to be made between different strategies, but choices will still have to be made.

A company's key performance factors are the answer to the question: "What do we have to be excellent at doing to get our potential customers' business?" This is a difficult question, because the only way to know if your answer is correct is to implement the strategy and monitor the results. Top management must be able to see the whole forest and choose which path to follow, instead of concentrating on the individual trees and going off in several directions at once. After that, management must separate what's important from what's not, and communicate that to the entire organization.

Defining strategic priorities is not a new problem, as illustrated by the following note received by the British War Office in London in 1812.

Gentlemen:

Whilst marching from Portugal to a position which commands the approach to Madrid and the French forces there, my officers have been diligently complying with your requests...

We have enumerated our saddles, bridles, tents and tent poles, and all matter of sundry items for which His Majesty's Government holds me accountable. I have dispatched reports on the character, wit, and spleen of every officer. Each item and every farthing has been accounted for with two regrettable exceptions...

Unfortunately the sum of one shilling and nine pence remains unaccounted for in one infantry battalion's petty cash, and there has been a hideous confusion as to the number of jars of raspberry jam issued to one cavalry regiment during a sandstorm in Western Spain...

This brings me to my present purpose, which is to request elucidation of my instructions from His Majesty's Government, so that I may better understand why I am dragging an army over these barren plains. I construe that perforce it must be one of two alternative duties, as given below. I shall pursue either one to the best of my ability, but I cannot do both.

1. *To train an army of uniformed British clerks in Spain for the benefit of the accountants and copy-boys in London, or perchance,*
2. *To see to it that the forces of Napoleon are driven out of Spain.*

Your most obedient servant,

Wellington

Like Wellington's bureaucrats back in London, many companies focus on the trivial, not on what matters most. They spend great effort to reduce direct labor costs because they are easy to measure, even though they may be only 10% of sales and are the least likely factor to increase a company's sales and profits. On the other hand, customer service, customer satisfaction, and quality performance throughout the company get no attention, even though quality costs are typically 20 to 30% of sales.

Determining key performance factors requires making choices and tradeoffs, because it is impossible to maximize everything. Making these choices is not easy, but it must be done. No company has the resources to be all things to all possible customers.

Deploying Key Performance Factors

Defining key performance factors identifies a company's principal business processes. The key factors then need to be reflected in the lower level process and department measures as shown by Figure 5-3. Cascading the key performance measures throughout a company may appear to be complicated, but it is not all that difficult, providing internal customer requirements are known and how processes work is understood. All those conditions may not be totally satisfied at the outset of implementing performance measures, but as will be discussed in Chapter 6, this does not prevent getting started.

Determining Process and Sub-Process Performance Measures

Assuming a company has a viable strategy, improving its total productivity depends on achieving and maintaining operational excellence. Superior

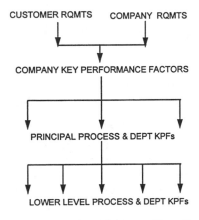

Figure 5-3   Deployment of Key Performance Factors

operations can compensate, to some extent, for a weak strategy. Poor execution, on the other hand, makes a weak strategy worse. It also makes it difficult to determine whether the fault lies in the strategy or its execution.

Process measures enable a company to control and improve its operational performance so when questions arise about strategy, the quality of operations is not a mystery. Determining process and sub-process performance measures requires understanding what a process is supposed to accomplish (what its customers want) and how it works. Then, it is a matter of identifying the variables that affect the quality of the processes' outputs and its productivity. Some proven techniques for accomplishing this follow.

Deployment Flowcharts

Preparing a detailed flowchart of a process is the first step that should be taken to understand how it works. A helpful variation on the standard flowchart is the deployment flowchart illustrated by Figure 5-4. In this format, the steps taken by different departments are arranged in columns so a horizontal line appears on the chart when work passes from one

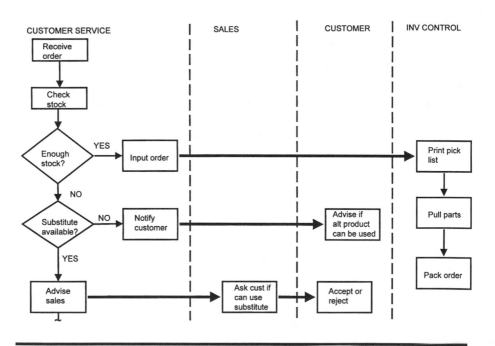

Figure 5-4   Deployment Flowchart

department to another. These lines indicate a vendor-customer interface, where quality problems are likely to occur that should be measured.

The flowchart will not say what should be measured, but its value goes far beyond indicating potential measurement points. Drawn to a proper level of detail, a deployment flowchart can reveal duplication, unnecessary steps, waste, and potential sources of problems that were not previously recognized. Most importantly, it will develop an in-depth understanding of how a process really works. As described earlier, this will often be far different from the way everyone thinks it works.

The key to developing a good flowchart is attention to detail. Every step, decision point, piece of paper, bit of data, and variation in how work is accomplished must be identified. It can take considerable effort to document a complicated process, but the insight gained is well worth the effort.

Cause-Effect Analysis

Another powerful analytical approach is to construct a cause-effect tree as shown by Figure 5-5. This is accomplished by starting with the broadest problem (or symptom) and breaking it down to successive levels of detail. The cause-effect tree is similar to the "fishbone" or Ishikawa diagram found in quality improvement literature, but I find that starting at the

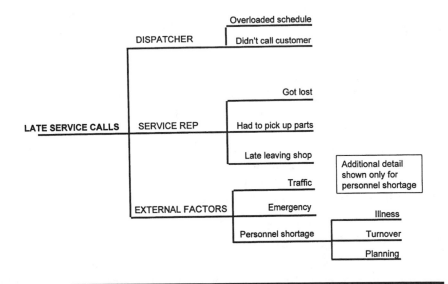

Figure 5-5   Cause-Effect Tree

highest level and working down through cause-effect relationships to successive levels of detail is an easier approach. This technique tends to make the user follow and understand the logical relationships of process variables, instead of grouping them under broad categories like materials and methods. Whatever technique someone favors is immaterial, what counts is understanding the cause-effect relationships in a process.

As in preparing a flowchart, getting down to details is the key to success. Ideally, the last branch of a cause-effect tree should be the specific causes of a problem. These are the most detailed variables that should be measured. In practice, the causes of problems may not be very well understood and getting down to particular steps in a process may be as far as you can initially go in determining cause-effect relationships. The ability to construct a cause-effect diagram is obviously limited by how well a process is understood. Consequently, developing one can be a demanding learning experience — but that is what makes it worthwhile. Normally, it will take several revisions to arrive at a reasonably complete diagram. Even then, additional cause-effect relationships and variables will probably be found later when the performance measures are used.

Strategic Deployment Analysis

Strategic objectives must be included in any performance measurement system. Strategic deployment analysis is a variation on the cause-effect tree that starts with the strategic objective and branches it backwards through all the company's operating units down to the department level as shown in Figure 5-3. It will duplicate much of what may be in the cause-effect analysis, but it provides an organizational, rather than a process perspective.

Strategic objectives not directly tied to operational performance will be identified by this analysis. For example, increasing production capacity to a certain level is not going to show up as a quality, waste, or productivity variable, but it can be measured. Another example, is a company that measures its use of the latest technology for producing its products, because it feels that it cannot afford to fall behind its competitors in its analytical and manufacturing capabilities.

Decision Analysis

Since managers have to make many decisions, one of the questions they should ask themselves is: "What performance measures and other information do I need to make the decisions I am faced with on a regular basis?"

The way information systems should be designed is shown by Figure 5-6. That is, the decisions to be made should determine the information needed to make them. This, in turn, should determine the data that must be collected. What often happens instead, is that the data created in the normal course of business is captured by an information system that then produces whatever information it can from the data. Managers then take the reports and use them as best they can to make the decisions they must make, even if the information isn't what they really need. Perhaps this explains why surveys I have conducted of middle managers from a wide variety of businesses indicate they feel they are getting only about half of the information they need to do their jobs.

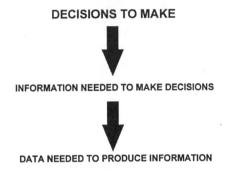

DECISIONS TO MAKE

INFORMATION NEEDED TO MAKE DECISIONS

DATA NEEDED TO PRODUCE INFORMATION

Figure 5-6   Decision Analysis

While all the causes of this information shortage are not readily apparent, at least part of the problem is caused by managers not assuming responsibility for determining what information they need. As Peter Drucker said:

> *"Few executives yet know how to ask: What information do I need to do my job? When do I need it? And from whom should I be getting it?"*[15]

These same surveys indicate performance measures are a large part of the information managers want, but are not getting. Managers want performance measures for two reasons:

1. To know what they are accountable for and how well they are performing
2. To know what is happening in their own areas of responsibility

Part of the solution to closing the information gap is for managers to take the initiative to get the information they need, instead of just making do with what they are given. The first step is to make a list of the decisions they face and the questions they need to answer on a regular basis. Then, opposite each decision or question, list the information and measures that will help provide the answer. Taking it another step, they should then take a blank sheet of paper and lay out the reports that would provide them with the information needed to make particular decisions.

Typical questions managers should be asking are

- Which processes are meeting performance objectives?
- Where are we getting better or worse?
- What should be my priorities?
- Where should we be applying our resources?
- Who do I need to see today? About what?
- Which managers/supervisors need help?
- Who deserves a pat on the back?
- Where do we need more people? Less?
- Where are the largest opportunities for improving performance in the process or department?
- What are the limiting factors or bottlenecks in the process?
- Where have we reached the limits of process capability?
- What looks like it might break down next?
- How is our workload changing?
- What additional resources will be required to accommodate an increased workload?

If all of the components of the production process measurement model are measured, most of the above questions will be addressed, but thinking in terms of specific decisions will identify any special measures that might be needed. Performance measures will not necessarily provide all of the information to answer any particular question, but they will definitely provide insight and understanding that cannot be obtained any other way.

For example, in a distribution company two very important questions were how well was the inventory management system working and where was it not working well? Measuring inventory turns (sales for a period divided by average inventory) and stockouts provided some insight, but these measures were not very sensitive and could hide many problems and opportunities.

Asking the question, "What is the inventory management system supposed to accomplish?" led to the answer that it was supposed to keep inventory of each item between specific limits. The limits were derived from sales, sales variability, the number of customers using the product, the ease of producing or purchasing the product, unit costs, where the product was in its life cycle, and other factors.

This led to the conclusion that a good measure of the system's performance would be how many days products were outside of their defined lower and upper limits. The key performance measure for the inventory system became:

$$\frac{\text{Total product–days out of limit}}{\text{Total product–days for the period}}$$

This proved to be a reliable and sensitive indicator. When applied to individual products, a summary report would look like Figure 5-7.

This measurement system provided good visibility into the performance of the inventory system and could identify products which had control problems such as products C, E, I and F. Looking at the behavior of individual products more closely, yielded another refinement. The data showed that some products seemed to be always near the upper limit,

Product	Days below (%)	Days above (%)	Days within (%)
C	36	2	62
E	31	0	69
A	12	0	88
H	8	3	89
B	6	8	86
G	5	2	93
D	2	0	98
J	0	12	88
I	0	27	73
F	0	37	63
Total	**13**	**9**	**78**

Figure 5-7   Inventory System Effectiveness Report

while others were always close to the lower limit. Further investigation revealed that always being close to either limit meant the product was easier or more difficult to procure than anticipated. In that case, it was appropriate to adjust the inventory control parameters or solve the procurement problem.

It was a simple matter to develop computer programs to identify products in either situation. This made it easy to identify potential problems and keep the control parameters consistent with real market and production conditions. The result was a 30% reduction of inventory while increasing on-time delivery from 86% to 94%.

This example illustrates the need to have detail in order to control a complex system. By being able to identify the individual products with too much or too little inventory, adjustments and corrections could be made to the inventory and procurement system on an individual product basis. The managers could not just see how the whole system was performing, they could quickly get to the actionable level (the individual product) and take corrective action.

Real-Time Quality Survey

A technique that is very effective in identifying quality problems in internal operations is the real-time quality survey. It is simply asking everyone involved in a process to record the problems they encounter in doing their job. A sample survey form is shown by Figure 5-8.

Instructions for using the form are

1. Ask everyone involved in a process to note any problems they encounter which they feel interfere with doing their job in the best possible way or the way they think it should be done. This would include any product or procedural quality problem that affects productivity, quality, or waste.
2. Problem descriptions should be brief and specific such as, "service request form not legible," "tool not sharpened," or "computer not working." Avoid using general descriptions such as "bad instructions" or "defective part." Say what is bad about the instructions or how the part is defective.
3. Use the "Tally" column, as shown, to mark how many times a particular problem was encountered and put the total in the "Count" column.
4. If applicable, the form can be modified to indicate the impact of the problem such as a minor/major classification or recording the actual amount of time lost or waste created.

QUALITY SURVEY Date _____

Name _Reggie Dinswaddle___ Dept/Operation _Shipping_____

QUALITY PROBLEM	TALLY	COUNT	NOTES
Instructions not clear on sales order	////	4	
Copier not working	/	1	Down for 40 minutes! Second time in 3 days.
Picklists not in order box	ᄊᄊ ///	13	
Wrong item pulled	ᄊ //	7	
Orders late from processing	/	1	Lost 35 minutes had to work OT

Figure 5-8   Quality Survey Form

5. Conduct the survey for a period that would probably capture most problems encountered in the normal course of business. Usually, 3 to 5 days is sufficient where there are many transactions each day. If it takes several weeks to produce a product, a longer sample period may be needed. A good guideline is to stop the survey as soon as it appears new problems are not being identified.

   In most organizations, it will require considerable follow-up by managers and supervisors to make sure the quality problems are recorded properly. Most people will have difficulty recognizing quality problems because they have become so used to seeing them, they think they are normal. Without management attention, there is a high probability that many of the sheets turned in will be blank. A good practice is to run the survey for one day, review all the survey sheets, provide appropriate feedback to everyone using some of the examples from the survey sheets, and then re-start the survey.

6. Review the survey sheets and make a list of all identified problems, giving them unique names. Supervisors or others familiar with the process should be involved in the review to recognize when different language is being used for the same problem.

7. Compile the list of specific problems into categories according to the responsible step in the process and rank them by the number of incidents. The problems that only happened a few times may be put in a "miscellaneous" category, but the others are repetitive quality problems that should be measured.

8. Ranking the problems by their total count (or other basis as in step 3) provides a very rough idea of their relative importance, but remember this is only a small sample that may contain many errors and is not likely to represent average performance.

Another way of accomplishing a real-time quality survey is to use a log sheet that travels with the work. This can be easier to use in some situations. A sample log for a manufacturing company is shown by Figure 5-9. A similar approach could be used in service operations as well. For example, someone installing and repairing burglar alarms could fill out a similar form, noting parts shortages, incomplete work orders, time delays, and any other problems encountered on jobs.

The real-time quality survey is a simple, but very effective, technique for identifying recurring process and product quality problems that should be measured. It is better than an after-the-fact survey because most people can't remember the details of incidents that happened more than a few days ago unless the incident was very significant. Although these surveys will identify some causes of problems, most of the "problems" listed will only be symptoms and some may be several levels removed from the actionable level. The urge to react to the survey or jump to conclusions about the causes of problems should be resisted until the performance measurement system becomes fully operational and more data is obtained.

Interviews

Interviews with managers, supervisors, and front-line employees can provide some insight into quality, productivity, and waste variables that should

JOB TRACKING & PROBLEM LOG

Job No _____ Qty _____ Part # _____ Desc_____

Customer _____

Release date _____ Time _____ By _____

Spec complete date _____ By _____

Job start date _____ Time _____ By _____

Step 1 Start _____ End _____ Reject _____ Rework _____ Other _____

Reason/problem _____

_____

Step 2 Start _____ End _____ Reject _____ Rework _____ Other _____

Reason/problem _____

_____

Step 3 Start _____ End _____ Reject _____ Rework _____ Other _____

Reason/problem _____

_____

.

.

Finish: Date _____ Time _____ By _____

Figure 5-9   Job Problem Log

be measured. This after-the-fact approach is limited by a person's memory of past events, so some problems may not be mentioned and any judgment of the relative importance of problems is likely to be way off the mark. Experience indicates that interviews will identify about half of the problems that would be identified in a real-time survey. Regardless, interviews provide a useful starting point. When interviews are conducted before doing a real-time survey, they can make the survey more effective by providing examples of what should be reported. Using these examples in pre-survey instruction will increase the understanding of the survey's purpose and the types of problems to be reported.

Asking managers what they think should be measured can identify some candidates for measurement. However, if the managers have not given their information needs much thought and their understanding of their production processes is limited, this is not likely to be very effective.

What should be measured depends upon what is important to the customer, the production process, and the company, not someone's opinion. This requires understanding what customers want, how processes work, and what the company must accomplish to be competitive. While interviews can provide some clues to what is important to measure, they can only provide a starting point. The rest requires serious investigation, analysis, and deliberation.

Work Input and Product Output Analysis

Measuring work inputs and product outputs is necessary to determine the work capacity of a process and to have measures of productivity. On the input side of a process, the interest is in knowing the demand for resources and the work input measure must be equal or proportional to that figure. On the output side, the product output measure must be equal or proportional to the value produced. In many cases, the name for the work input and product output will be identical, but the same name really represents two different variables. For example, a painting process could have frames to paint (work) and produce painted frames (value).

Determining the work coming into a production process may seem to be quite simple, but it is easy to see the major work inputs and overlook the many little pieces that add up to a significant portion of the total workload. This is especially true of extra work created by quality problems, which can easily consume 20 to 30% of an operation's resources.

All work is driven by transactions. There is a cost attached to every piece of paper handled, telephone call made, and memo written. This is especially important for administrative functions, where small tasks can make up a large portion of the workload. When work inputs are not measured, "overhead creep" can occur because each new task is seen as a trivial addition that can be absorbed by the department. That may be true in the short-run, if there is excess capacity available, but it is never true in the long run.

To measure work input, it is necessary to account for all the major work elements, but it is not necessary to record every transaction. Keeping detailed records of all work coming into a unit for 5 to 10 days will usually be sufficient to identify the primary drivers of activity and provide a useful ranking of their importance. Summarizing these as shown by Figure 5-10 will also provide a gross measure of non-productive hours in a department. Rework from internal or external sources (complaints and returns in Figure 5-10) should be in separate categories since it may be productive from a departmental viewpoint, but is non-productive from a company viewpoint.

Once the primary sources of work have been identified, they do not have to be measured every day or week. Where this data can be easily captured, it makes sense to do so, but where manual methods are necessary, sampling can be used. Taking a one-week sample every few months would be sufficient to monitor the workload, except in rapidly changing conditions.

Task	Avg. time req'd per unit (min.)	Units in sample period	Core time* required
Inquiry – product availability	2.0	611	1222
Inquiry – order status	1.7	556	945
Orders – telephone	4.1	2204	9036
Orders – mail	3.2	1003	3210
Complaints	6.0	96	576
Returns	4.3	187	804
End-of-day verification checks	135.0	5 days	675
Weekly Total			**16,468**

* Core time = the normal amount of time to do the work, determined by a work sample study.

Minutes available for the period	21,600 (9 people × 40 hrs × 60 min.)
Less core time required	(16,468)
Non-productive minutes	**5,132 - 24%**

(Wasted, spent in meetings, correcting problems, etc.)

Figure 5-10 Department Activity Summary

A similar approach can be taken to identifying product output, but care must be taken to identify the right "product." Physical products are readily identified, but identifying the outputs of services can be tricky. The result the customer wants must not be confused with the activities to fulfill those wants. If telemarketing calls are supposed to produce qualified prospects, the product is qualified prospects, not phone calls. Activities are rarely valid products, it is the result or impact of an activity that customers want.

What products does a training department produce? Hours of classes? Increased skills in an organization? How about fewer complaints as a result of the improved skills? The final product is a reduction in complaints, but the hours of classes and the change in skills should also be measured as intermediate products to understand the total process.

When customers of a process and their requirements are known, identifying service products should not be difficult. If customers of a service can't be identified or they don't know what they want, the service should be discontinued because there is no value to be added in the first place.

A company that wanted to reduce overhead costs, took the approach of asking the various operating units what services they wanted to buy from the administrative departments. This seemed like it would work — until no one wanted to buy about half of the services provided by the human resources department.

This was a clear indication that the operating units felt they weren't getting much value from the services being provided. The services should have been eliminated or modified so they would meet the customers' requirements. Instead, a token human sacrifice was made of one poor soul in the department and the operating units were strongly urged by the CEO to buy the department's services. Faced with an offer they could not refuse, the department's services were quickly "bought" and life went on as usual, without solving the problem.

No business function or department should just "do" — it should do with a purpose. That purpose must be to provide its customers with products that provide value and meet their requirements.

Composite Measures

If a process has many different types of inputs or outputs, they need to be combined into an aggregate, or composite, measure. A good technique for constructing an aggregate input or output measure is to reference all products to a standard product that is familiar to everyone.

Consider an insurance company that processes applications and issues new policies, renews existing policies, and cancels policies that have lapsed for non-payment of premiums. Each of these tasks has an average work content that could be determined. These figures could then be used to reference everything back to a "renewal policy" equivalent as shown by Figure 5-11.

	Quantity (A)	Standard Hours/unit (B)	Standard hours (A × B)	Renewal policy units (Std hrs/ 1.2 hrs/unit)
New policies	430	2.9	1247	1039
Renewals	1000	1.2	1200	1000
Cancellations	210	1.6	336	280
Total			**2783**	**2319**

Figure 5-11   Calculation of Work Input as Equivalent Units

In this example, the standard hours (quantity produced X standard hours/unit) would be an equally valid measure of input, but using a familiar physical unit as a reference makes the measure more meaningful to people in the work group.

Measuring Productivity

Productivity is defined as output (value) divided by resources consumed. Simple in concept, but difficult to apply in very broad processes because there can be many different types of resources consumed and outputs produced. Here, financial ratios such as sales dollars to process costs can provide gross indicators of productivity, but such measures are not going to be very sensitive to changes in performance. For example, any productivity measure based on sales may, in the short-term, reflect changes in sales more than real changes in performance.

A more meaningful approach to measuring productivity is to concentrate on separate processes or functions. Here, the outputs are fewer in number and their relative value should not be difficult to determine. For example, gross measures of accounting productivity could be derived from financial ratios, but if accounting is broken down into accounts receivable, accounts payable, taxes, and other processes, measuring productivity in each process becomes more manageable. Separate productivity and quality measures of each function could be developed which would be more meaningful and more sensitive to change. For the purposes of monitoring and improving performance, productivity measurement efforts should be concentrated on individual processes producing a narrow range of products. If the productivity of the parts is taken care of, the whole will take care of itself. However, aggregate productivity measures should also be used to verify the whole is reflecting the parts.

Both outputs and resources consumed can be expressed in terms of physical units or financial terms (price or cost). Wheelbarrows/labor hour, dollars of wheelbarrows produced /labor hour, wheelbarrows/labor dollar, and dollars of wheelbarrows produced/labor dollar would be equally valid measures of direct labor productivity of a wheelbarrow plant.

Waste Analysis

Waste is any resource consumed which does not add value to the product. Rework and scrap are obvious forms of waste, but there are more subtle forms that should be identified and evaluated to determine if they are significant enough to measure.

- Downtime — when some link in the process fails to operate
- Idle time — time lost waiting because work is not available
- Queue or delay time — time work spends waiting for the next step to take place
- Inventory — finished goods or work-in-process
- Under-utilized capacity — capital providing no return
- Movement — unnecessary movement of materials, work, or people
- Motion — more than necessary, while performing work
- Performing any task that adds no value

In most operating environments, the recurring forms of waste that should be measured are rework, scrap, idle time, downtime, and queue time. Under-utilized capacity and excessive inventory, movement, and motion are more structural in nature and should be addressed by observation, mapping, and other analytical techniques. Improving operational performance will relieve these structural problems to some degree, but that usually won't address much of a structural problem.

Determining Relative Importance

To establish priorities, the relative impact or importance of quality problems must be determined. For internal quality problems, measuring the frequency of occurrence is not sufficient. The best measure is costs. It is not very difficult to estimate the cost of rework, scrap, downtime, idle time, or inventory. Determining the cost of queue or delay time is a less tangible problem, but for a specific process, it is possible to identify the effects of such delays. For example, the additional cost of overtime to make up the time lost could be estimated and used as the cost of delay time.

The cost figures that will be developed may only be rough estimates, but if the same people make them using consistent ground rules, they will be a reliable indicator of the relative cost of quality problems. They will also probably not be terribly far off the true costs.

An important consideration in calculating the cost of rework and scrap is where a problem is discovered in the production process. Every step in a process is supposed to add value to the product. This increase in value, as illustrated by Figure 5-12, means that the cost of a defect found at the end of a process is going to be much greater than finding the same defect at the beginning of the process. The actual cost of any particular

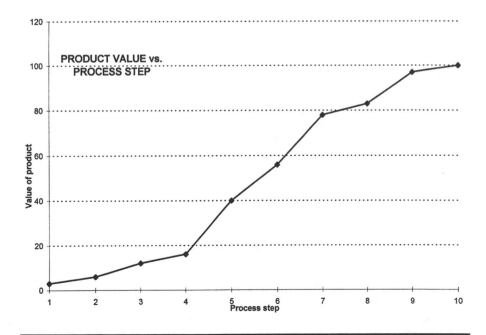

Figure 5-12   Value-Added Profile of a Production Process

defect would also depend on the nature of the defect, its severity, and other product or process variables.

Any quality problem that gets to a customer must have a higher cost than just correcting the problem. Consequently, for external customers, two quality costs must be added to determine the true cost of a quality problem:

1. All the tangible costs of repair or replacement
2. The less tangible, but higher costs of the impact on customer satisfaction — increased selling costs, lost sales, and lost customers

Putting a cost figure on customer dissatisfaction is not easy to do. In some situations, it might be zero; in others, a few problems could result in the loss of an important customer. Specific situations might be easier to quantify, but where a physical product is concerned, multiplying the actual cost of repair or replacement by a factor of five to ten is very reasonable. The basis for saying this, is an analysis of quality costs by one authority on the subject, which led him to conclude that the long term cost of quality is at least six times what anyone can identify as tangible costs.[16]

For services, there is also a rework cost that can be estimated. Costs should be estimated as closely as possible, but for establishing priorities, what is most important is the relative size of costs or opportunities, not their precise value. If the method of valuation is consistently applied, a 20% error is not going to make any significant difference in priorities.

Costs or opportunities can be expressed as dollars (or some other currency), but it could take considerable time and effort to put cost figures on everything, especially if the necessary financial data is not available. An alternative is to use a weighting scheme and assign points to quality problems according to their estimated actual costs. For example, assume a process has four steps and within the process, three different degrees of defect severity are possible:

1. Minor, requiring less than ten minutes work to correct.
2. Major — requiring an average of one hour to repair.
3. Reject — the product must be scrapped.

Points could be assigned to each category of defect, but as the product goes through each process step, the cost of each defect is going to increase because:

- More value is added to the product.
- The time required to correct problems is going to increase.
- There is a greater probability the repair will not be successful.

Step	POINTS		
	Minor	Major	Reject/scrap
1	1	5	10
2	2	10	25
3	5	25	50
4	10	50	100

Figure 5-13   Point Matrix: Process Step and Defect Severity

A table of weights assigned to each combination of defect category and process step might look like Figure 5-13.

The specific type of defect and its location on the product could also have a significant effect on its cost. If so, that would add two more dimensions of complexity to the matrix. This level of complexity may seem difficult to handle, but it can be easily accommodated by relational database management systems. There is no reason to shy away from using the degree of complexity necessary to reflect what happens in a production process.

As experience is gained and the impact of quality problems becomes better understood, point weights can be adjusted. Then, when the economics of the production process and the costs of customer dissatisfaction are determined, a cost per point figure can be calculated based on an estimate of total quality costs and the total quality points created for the same period.

With both goods and services, value and quality costs increase with each successive step of a process. Performance measures must reflect these cost relationships, but general rules cannot be applied — it all depends on the production process and the customer. Using point weights to establish the cost relativity of quality problems, is a way to get meaningful performance measures implemented without having to do an exhaustive study of costs.

Measuring Variables the Right Way

As Dilbert shows, it is possible to create performance measures that are meaningless nonsense instead of useful information by measuring things the wrong way. An example of this is a company that thought it was providing excellent customer service because its on-time-delivery rate was

DILBERT reprinted by permission of United Feature Syndicate, Inc.

greater than 99%. The only problem, was that "on time" was defined as being delivered when delivery was promised, not when customers wanted the product. In addition, the "promised delivery date" was only promised when it was certain the product would be available. To make matters worse, the delivery date represented when the product was shipped, not when it was actually delivered. The only way this measurement could have been more biased in the company's favor was if customers entered the promised delivery date when they finally received the product. The actual on-time-delivery rate was 70%, not 99%.

The way on-time delivery should be measured, is to measure the difference between when customers want a product and when they receive it, but few of the order entry systems I have seen provide for capturing this information. They also don't usually capture orders lost because products were not available or requested delivery dates could not be satisfied.

Subtle differences in performance measures can make a significant difference in what performance appears to be, as well as in customer satisfaction. For example, how should a catalog retailer count orders being shipped on time? Only when all items on an order are shipped on time or should each line item be counted as a separate "order?" It will make a big difference in the measurement of "percent of on-time shipments". Only customers can answer this question. Deciding what to measure is the largest question managers must face, but deciding how to measure a performance variable requires taking the customers' point of view and giving some careful consideration to what is really being measured.

Measuring the wrong things the right way can also happen. When determining what to measure, it is easy to make a snap judgment that seems correct, but which may have some unintended consequences. Putting emphasis on productivity, for example, will encourage a manager to keep a plant operating at full capacity by building finished goods inventory. Is maximizing productivity a rational goal? What should the plant have as its primary objective? Maybe it should be to convert raw materials to cash as quickly as possible, while striving to do so at minimum cost.

Minimizing acquisition and service costs per customer is another conventional business objective. But minimum costs don't mean maximum profits. If it did, all companies should go out of business since this would bring their costs down to zero. Peter Drucker argues that banks should not focus on cost per customer but the yield per customer (volume and mix of services a customer uses), since this determines costs and profitability.[17]

The question that must always be kept in mind is: "What is this process supposed to accomplish?" The answer to that question should be the primary or key performance measures for the process and everything else should be secondary or lower in importance. If that question cannot be answered, additional investigation, analysis, and deliberation must take place until it can be answered.

Measuring the Unmeasurable

Putting numbers or values on subjective factors such as appearance, the quality of a report, or level of dissatisfaction is a common measurement problem. As mentioned in Chapter 2, this seemingly impossible task is accomplished every day in sports such as gymnastics and ice skating as well as in judging everything from cows to choirs — and it is accomplished with a good degree of consistency.

Close examination of how this is accomplished shows that there is a well-defined quality "standard" that is used as the basis for comparison. This standard may not ever be actually achieved, but it is defined by many specific items that can be seen, heard, or sensed in some manner. What blade edge an ice-skater lands on, how a gymnast turns, or where a diver's toes are pointing, are all clearly defined specifications for the "product" — the performance during the competition.

Once the specification has been defined, judges (inspectors) must be found or trained who understand the specification and have the skills and knowledge to compare the actual performance against the standard. That this can be done with a very useful degree of accuracy and repeatability has been demonstrated countless times in many fields. Qualified panels of experts have been used to measure everything from the taste of wine to the odor of materials used in space vehicles.

Similar techniques have been used to measure subjective factors such as the level of customer satisfaction after a complaint has been resolved.

Satisfaction index	Customer's response
10	Sends flowers or writes "thank you" letter to CEO.
9	Writes "thank you" note to service representative.
8	Expresses satisfaction in specific terms.
7	Says "very good" or similar positive comment.
6	Accepts solution, says "thank you."
5	Accepts solution with neutral or no feedback.
4	Acts irritated. Curtly says good-bye.
3	Expresses dissatisfaction in specific terms.
2	Calls service representative nasty names.
1	Threatens to sue company.

Figure 5-14   Numerical Scale for Quantifying Customer Satisfaction With Company Response To Complaint

In applications such as this, an operational definition of how the level of satisfaction is to be determined is required. In this case, a scale like the one given by Figure 5-14 might be used.

Measuring the unmeasurable, requires translating a vague concept into specific requirements. Those requirements must then be converted into operational definitions that can be practically applied to make judgments with a high, but not necessarily perfect, degree of accuracy.

SUMMARY

Determining what performance variables to measure in a production process, a department, or a company, is part science and part art. The science says all of the components of the production process model should be measured and those measurements should reflect all customers' needs and wants — including the company's own interests. The art side of the question says that because no company has the resources to be all things to all people, choices must be made about which customers to serve and how to compete for their business. A strategy is the result of making those choices and that strategy must be reflected in the performance measures.

The procedures given can identify most of the variables that should be measured, assuming customer requirements, a company's strategy, and its production processes are well understood. Since this is not usually the case, implementing performance measures is a process of learning and development, which will bring these questions and their answers to the surface.

Determining what to measure does not end with compiling a list of all process variables that must be measured. Their relative importance to customers, the company, and production processes must be determined along with exactly how each variable will be measured. Determining what to measure is completed when the process variables, their relationships, and how they will be measured has been determined. What is incorrect with some of these conclusions will be found by implementing and using the performance measures.

6

IMPLEMENTING PERFORMANCE MEASURES

The first decision to be made when implementing performance measures is where to start. Ideally, implementation should begin with identifying external customer groups and determining what they want. Then, a strategy should be developed which will lead to defining the company's key deliverables, performance factors, and production processes. Measures should then be implemented in the principal processes according to where the greatest return or risk exists.

While this sequence is conceptually sound, it is not always necessary to have a well-defined strategy before implementing performance measures. Since everybody wants whatever they are buying cheaper, better, and faster, there are usually significant areas of opportunity available that can be addressed with performance measures while any strategic questions are being resolved. When costs are too high, deliveries are chronically late, or customers are complaining about product quality, it doesn't take a detailed long-range strategic plan to figure out what to work on first. Furthermore, when a company's operations are not performing well, its competitive position may be so unclear that trying to develop a strategy would be very difficult.

> *In a conversation with a sales manager for a large furniture manufacturer, I asked him what product quality factors could be improved that would give the company a competitive edge. Expecting to hear about some finer points of quality, I was stunned when he replied: "Well, it would certainly be a big step forward if we could deliver what customers ordered and get it to them on time... No dents or scratches would also be a big help."*

Getting performance measures in place will identify where changes in strategy should be considered. For example, when performance measures were implemented in a graphic arts company, they showed rush orders accounted for most of the rework and waste in manufacturing and added significant costs to support departments. When all these costs were properly allocated to rush orders, it became apparent they were not profitable. The company then had to choose between continuing to lose money on rush orders, not accept them, raise prices, or improve its internal procedures so rush orders would be profitable.

There are no universal answers to the question of where to start implementing performance measures. It all depends on a company's situation and where it can expect to get the greatest return. In any case, attempting to implement performance measures simultaneously throughout a company is probably not the best approach, since this can overload management resources.

A better approach is to start with a few pilot areas where there are some worthwhile opportunities. In addition to providing the greatest return on investment, this approach enables resources to be concentrated, decreasing the time it takes to develop a working system. Then, when the measures lead to some tangible improvements in performance, the measurement system's credibility will be established. Since getting results is the best way to promote change,[18] achieving measurable improvements in performance will also get most of the nay-sayers standing on the sidelines to buy into performance measurement and process improvement.

Pilot projects also provide a valuable learning experience for the people doing the implementation. When developing a performance measurement system, any number of technical and organizational problems can be expected. It is far better to test the waters before diving headfirst into the pool. Little, if any, time will be lost by using a pilot project approach because the knowledge gained in the pilot areas will make subsequent implementation much easier.

It is also generally better to start at the end of a process than the beginning, because the tail of a dog is where all the problems in the rest of the animal will show up. In the manufacturing process shown by Figure 6-1, performance measures in the assembly department will identify most of the problems created for manufacturing by sales, customers, engineering, parts manufacturing, purchasing/vendors, and inventory control, as well as those caused by the assembly department itself.

Once the assembly department measures are established, they will point to the functions that are the major causes of problems in assembly. Logically, these would be the next areas in which to implement performance measures. Starting at the end of a process and working backwards

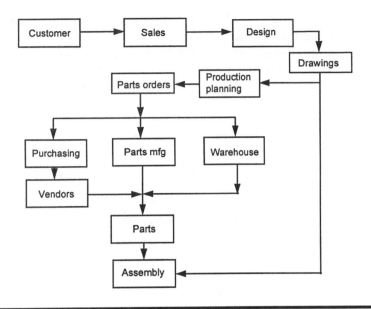

Figure 6-1   Manufacturing Process

follows the paths of most opportunity through the cause-effect diagram of a production process. These paths will cross departmental lines, so some gaps may exist in the performance measures until the complete system is implemented. However, this approach will provide the greatest return on investment.

Laying the Foundation

After the places to start implementation have been selected, the next step is to secure organizational support for the program. Everyone involved needs to be told why performance measures are being implemented, how the measures will be used, and what everyone is expected to contribute. When people hear the words "performance measurement," their emotional response will run the full range from absolute fear to optimistic enthusiasm. The most prevalent attitude among employees and managers alike will be: "It might work, but I'll believe it when I see it. I'll go along with it as long as I have nothing to lose." This is perfectly acceptable. After all, who can find fault with anyone for having reservations about something new until they see it work?

Implementing performance measures can be more threatening to managers than employees. It raises some difficult questions about strategy, priorities, and accountability that must be answered. In addition, when unrecognized or deliberately buried problems start coming to the surface, it can be upsetting and embarrassing.

The primary questions everyone will have about performance measures, along with appropriate replies, are given below. If a company's managers disagree in principle with these answers, then some adjustment in their attitude is required.

1. **Why do we need performance measures?** — To know whether our performance is getting better or worse and to identify where we need to improve to better serve our customers and operate more efficiently.

2. **How will the measures be used?** — After the data is collected, performance will be summarized for each responsible department, with the problems ranked by impact or cost. Then, we will select problems or opportunities to attack, according to where we think we can get the most gain for the least cost and effort. After assigning specific items to managers or teams, the measures will be used to determine if performance is really improved by the changes made.

3. **Will they be used to find fault and punish someone (especially me)?** — No one is going to reprimanded or fired for making a mistake. None of us is perfect and we know the way we do things can be improved. If someone is guilty of gross negligence or keeps making the same mistake over and over, that is a different matter. However, performance measures have never been needed to identify that kind of problem anyway.

4. **Why should my performance be measured?** — The focus is not on a person's performance but on problems and opportunities in our operations and processes. It is true, however, that the performance of departments, and even some individuals, will be measured by the system. We feel that everyone's performance should be measured and that everyone should be accountable for performance in their area of responsibility. That has always been true for the president and other officers and managers of the company. In that sense, all that's happening is that performance measurement is being refined and expanded to include everyone.

5. **What's in this for me?** — Like all companies, if we don't serve our customers well and operate more efficiently, the Frammis & Glotzen Company will cease to exist. So the first issue is survival. Beyond that, if sales and profits increase, everybody will benefit by better wages and benefits, more opportunities for advancement, and more job security. Companies that don't make profits, don't give raises, promotions, or benefits, because they can't afford it.

6. **Why are we being picked on? Why is this starting in my department/division?** — This is a company-wide program, but we don't have the resources to implement this program in all areas of the company simultaneously. You have to start someplace and your unit was selected as one of the pilot areas because we think there are some significant opportunities here and your group can make it work. What we learn from these pilot projects will make it easier to implement similar performance measures in other areas.

7. **Will we see the measures?** — Yes. The measures that apply to your department will be posted or available in your area. The top-level measures for each department, function, or process will also be displayed on a bulletin board. The detail reports will also be available to anyone who wants to see them. There will be no secrets, but in a few cases, performance measures that must be kept confidential for competitive reasons will use indexes, rather than the actual numbers.

Some training in the principles of quality and productivity improvement and problem solving may be needed so that everyone will understand why performance measures are necessary. This training doesn't have to make everyone an expert in either subject. All that is initially needed is to create a conceptual understanding of how the measures will be used to improve performance.

If there are no signs of organizational unrest or violent objection to performance measurement after communicating and training, implementation can begin. Otherwise, additional counseling and training may be required. It may even be necessary for management to make some guarantees about the measures not being used for punishment. However, I have never experienced a situation where getting the cooperation of front line employees was a problem if management provided the necessary support.

Determining What to Measure

With the processes selected and the organization primed for implementation, determining what to measure can begin. For the selected processes, the first step is to determine the primary deliverables and their key performance factors. Some key factors may be readily apparent, but it is not a good idea to jump to conclusions without getting some feedback from customers. When companies assume they know what their customers want, they can overlook quality issues that customers value very highly.

After the key factors from the customer's viewpoint have been identified, determine what is most important to the company. There may be waste factors which have a large impact on cost or there may be strategic issues which are especially important. In one high-tech company, for example, the capability of one manufacturing process is one of its primary competitive advantages.

Once the key performance factors have been identified for the processes being addressed, the techniques described in Chapter 5 can be used to identify internal process measures. The most efficient approach, is to use the different methods for identifying what to measure in the following sequence:

1. Prepare deployment flowcharts of the process and identify the critical interfaces in the process.
2. Interview managers, supervisors, and front-line employees about the problems they see or encounter in the normal course of business. This will identify some performance or quality issues that can be used as examples of the type of problems that should be entered on the quality survey form.
3. Conduct real-time quality surveys to identify internal process quality problems.
4. Construct cause-effect diagrams of the process, so the relationship of the problems identified by the surveys is understood.
5. Use decision analysis to define the reports and charts that would provide managers with the measures and other information they need. Compare this with the outputs of steps 1 to 4 to see what other data will be needed to produce the reports.

Designing the Data Collection System

After these steps have been completed, what to measure and the data needed to provide the performance measures will be reasonably well defined. The next step is to start collecting data. This requires a data collection system to be designed and implemented. How this can be accomplished depends on what has to be collected and what mechanisms are already in place. Most companies already have some measures of resources consumed and outputs produced, or at least the raw data. Some of the higher level key performance factors may also be available from current sources.

What is most likely to be missing, is the detail data about quality and waste problems. A system will have to be designed and implemented to capture that data. It is probably best to start with a manual data collection system, because at this point, there will be many unknowns about what data is required. After the system has been used and the system design has stabilized, data collection techniques such as bar-coding can be used. Of course, if these data collection systems already exist, there is no reason not to use them if adequate support can be provided to the project.

Some samples of quality codes and forms for manually recording data are provided in the case study of Appendix B. As can be seen from the list of quality problems given by Figure 4 of Appendix B, there are many things that can go wrong in even a well-structured and mature process. It might seem like having to deal with a hundred different quality codes would be difficult, but any one person is going to see only a small portion of all possible problems. People are generally only going to see the problems that affect them. In most cases, that is all they should report anyway, to avoid duplication. Of a hundred different problems that may exist in a process, any one person will probably only have to deal with fewer than a dozen. For this reason, reporting becomes almost a reflex action after a few weeks, even though there will initially be some confusion about what to report and how to do it.

There is no way to give any guidelines for how many quality problems can be experienced in a process. That depends on the complexity of the process, which is determined by the number of steps in a process, the opportunities for quality problems to occur in each step, the external factors that affect the process, and the variability of its inputs and outputs. In a simple process, there might be only a handful of specific quality problems. In a complicated process, there might be a few hundred. Most processes are much more complicated than they first appear. For example, 22 variables were required to properly measure one company's rather ordinary order entry process.

The first problem that will be encountered after designing a data collection system, will be getting everyone to report reasonably accurate data. This will be a challenge, no matter how much instruction is provided. Most people get into the routine very quickly, but some will require considerable support and instruction. Intensive follow-up and checking of detail is needed to be sure all problems are reported and that they are reported correctly. Having all the blanks on a form filled in doesn't mean an incident has been accurately reported. This is where the responsible supervisors or department heads must get involved. During the first several weeks of implementation, they should review all data sheets and look into any entries that look questionable. If bar-coding or some other electronic data collection method is used, the data can still be verified and corrected before it enters the system.

If the performance data is incorrect, everything that follows is worthless, so training and intensive checking of data will pay big dividends in the future. This can initially significantly add to a supervisor's workload, but after a few weeks, most of this will no longer be necessary.

Implementing a performance measurement is similar to creating a new language. Quality problems, process steps, and other variables must be given unique names that have the same meaning to everyone. Then, everyone has to learn how to speak the new language. This can be difficult and frustrating, but there are no shortcuts. When a data collection system is implemented, many changes will be necessary to the data collected and any forms or procedures for collecting data. Most of the changes will be minor, but they will make the difference between success and failure of the system.

Developing the Data Processing System

After the data is collected, it must be converted into useful information in the form of reports and graphs. In all but the very simplest of cases, a computer will be needed to manipulate the data. The best approach is to use a relational database system. These software systems make it easy to conduct analyses, construct special reports, and make changes to the measurement system. Any of the current popular relational database systems is capable of handling the job. It is important to note that the logical relationships for establishing accountability for quality problems can be somewhat complicated. These and other logical operations may be beyond the capabilities of query and report functions built into some database systems, so some special programming could be required.

In the early stages of implementing performance measures, many changes will have to be made to the data collection and processing systems. Everything from what data is collected to what is on the reports will probably change during the first few months of a measurement system's life. It is very important that necessary changes are promptly accomplished. If this is not done, the credibility of management and the measurement system will rapidly deteriorate. If it appears performance measurement is not important to management, it will certainly not be important to front-line employees.

Since many changes will be necessary to the data processing system, the best approach seems to be to develop an unsophisticated system that works rather than an elegant system that takes forever to design and implement. For example, the computer systems group of a large company put together a comprehensive customer service measurement system in less than three months. It was not fully documented, contained many plugs and patches, and had some rough edges, but it worked. Then, knowing what was needed in the system, they went back and built an entirely new system that eliminated all the shortcomings of the initial system. According to the project leader, this "design it as you fly it" approach was very efficient and probably reduced the cost of getting to the final system.

What must be kept in mind is that good measurement systems are developed, not designed on paper and then installed. Going where no one has gone before, is going to lead to some missteps and backtracking. Anyone who claims to be able to get it all correct the first time is missing a few cards from his deck.

Developing the System

There is only one way to develop an effective performance measurement system — use it for all of its intended applications, especially for improving process quality. The first cut at a measurement system will probably be about 75% correct. That is, 75% of the data and information produced will be reliable and useful. The remainder will be inaccurate, worthless, or missing.

This figure must be brought up to 95 to 98% by trying to use the information and discovering what doesn't make sense, is wrong, is missing, in the wrong format, given to the wrong people, too detailed, not detailed enough, not in the right order, and so on. There are no quick cures or other alternatives to just grinding out the problems one by one.

Most of the changes required will have their roots in not understanding how a particular process works, rather than organizational or computer system problems. This gets back to the principle that if you aren't measuring a process, you can't understand how it works. But the catch is, if you don't understand how a process works, how can you measure it? The answer is that you start with what you know (or think you know) and learn as you go. Hopefully, you will go around in ever-decreasing circles until the point is reached where everything works and the performance measures are providing accurate, useful information.

Validation

Validation means that the people using the system, and measured by the system, accept the measures as presenting a generally accurate and reliable picture of performance. This condition can only be reached through training, using the measures, answering any questions that arise, and correcting any problems with the performance measurement system.

When performance measures are first implemented, there will be many questions about what they mean, where they come from, and how they should be interpreted, even if all this was explained many times before starting. Managers and the system developers should carefully listen to any questions or objections, because they may indicate where the system needs to be improved. Even if that is not the case, all questions must be answered. It is foolish to expect anyone to use or respond to something they don't understand.

It is very important in the early stages of implementation to verify the performance measures are being used. If the outputs of the measurement system are not being used, there are only four possible explanations:

1. There is a lack of leadership of the program.
2. The users don't understand the information.
3. The information is not relevant to the users' needs.
4. The information is incorrect or unreliable.

In any case, the problem needs to be identified and eliminated.

During the first several weeks, there will be many questions about the meaning of reports, charts, and related terms. The accuracy and validity of the information will also be questioned. These questions must be answered, not ignored or dismissed.

Getting a measurement system working and accepted can take several months in situations where there are many (at least a few hundred) transactions each day. In slower moving processes it may take longer than that. That does not mean, however, that the measures are useless until full acceptance is achieved. Many improvements in productivity and quality can be achieved with measures that still have some rough edges and are not universally embraced. In a ranking of quality problems, for example, the exact position of the top 3 to 5 problems may change in subsequent weeks, but they will probably all still be significant problems.

Getting the first week's reports will be similar to having one point on a graph — it doesn't provide a lot of information and may trigger more questions than answers. With a few more weeks, priorities for recurring problems will become clearer. In another 6 to 8 weeks, patterns may start to appear and priorities should be readily apparent, at least from a cost or impact perspective.

How do you know when performance measures have been accepted by an organization? Fireworks or other signs are not going to appear in the sky, but some reliable indications are

- Questions about what the measures mean and complaints about their validity disappear, for all practical purposes.
- Managers and supervisors are requesting changes or additions to reports and graphs. This indicates the information is being used.
- If the graphs and reports are late for any reason, managers and supervisors call to ask when they are going to get them.

Refinement

Even if the performance measures have been accepted by everyone concerned, there can still be many areas for improving the measurement system. For example, knowing exactly what data is needed may enable the use of special forms or input devices to reduce the number of manual inputs and chances for error. Another possibility is revising reports to make them more useful or eliminating reports that seemed like a good idea at one time, but are never used. Getting input from all users about their likes and dislikes can identify opportunities for improving the measurement system's efficiency and effectiveness.

Anything within reason that can be done to make the system more efficient and reliable should be done. The two most important factors affecting a measurement system's effectiveness are how relevant the information is and how easy the system is to use. The users must contribute by providing feedback about the value and usefulness of the information they receive, but the system developers bear most of the responsibility for making it easy to use.

SYSTEM DESIGN CONSIDERATIONS

There are many ways to collect data and convert it to useful information. How this can best be accomplished in a company depends on what systems it already has in place, its human resource capabilities, and its available resources. There are, however, some general guidelines which come from lessons learned through experience. They may not prove to be the best advice in every situation, but until there are good reasons to deviate from them, they should be followed.

Make it Easy to Report Data

Make it as easy as possible to record or enter data. Making someone walk 50 feet to punch in numbers on a computer terminal will only provide a reason for not reporting anything. Ask people to report only the data that is needed and put the burden of dealing with process complexity on the computer system, not on the people. In that regard, it is strongly suggested that ex-Internal Revenue Service employees not be hired to design data forms or write system instruction manuals.

Don't add steps and people to a production process to capture the necessary data. Build reporting into the process by modifying forms and procedures. Most people will be reporting 10 to 30 incidents per day, not hundreds. If it takes 10 to 20 seconds per entry, it will add 5 to 10 minutes to the daily workload. However, complaints about the reporting burden should be taken seriously and investigated. Most complaints disappear when everyone becomes more familiar with the procedures, but some will be legitimate and require modifying the system.

A few employees may argue that they weren't hired to report data, but to sell, answer calls, run a machine, or whatever. In that case, it may be necessary to point out that all job descriptions have been revised to make reporting part of everyone's job and not reporting is not an option.

Use a Rifle, Not a Shotgun to Hit The Target

Don't take the approach of collecting every bit of available data and rearranging it into massive reports no one can use. Instead, first determine what information is needed and then develop the system to supply it.

Reports and graphs should be designed for specific users and purposes. Since different managers must make different decisions, they need different information. Give everyone the graphs and summary reports of all measures that are relevant to them, but don't give them anything that is not relevant to them. All performance information should be available to everyone, but that does not mean they need to get every report produced by the system.

Use a branching relationship from the top to the bottom of the organization or through the cause-effect relationships of the process, to summarize and distribute performance information. As shown in Figure 6-2, this will mean a typical manager or supervisor will only have to monitor about 5 to 10 variables on a regular basis. If a person needs additional detail to understand a change in a higher level measure, it is then a simple matter to drill down to lower levels.

While measurement is process oriented, performance must be summarized by department, work group, or function to establish accountability. For other analytical purposes, summarizing performance by each step in the process and by other dimensions described in Chapter 7, usually provides the most relevant information.

> **President** – monitors company's key performance factors (KPF)
> and KPFs for each of five department VPs

> **Each VP** – monitors their unit's KPFs and KPFs
> for each of 3-6 department managers

> **Each department manager** – monitors
> their department's KPFs (1-3) and their
> quality problems ranked by cost or
> impact

Figure 6-2   Report Distribution Structure

Decentralize the Measurement System

Don't try to build a centrally controlled, one-size-fits-all system. It will be cumbersome, slow, and inefficient. In this age of powerful personal computers and computer networks, there is no reason to centralize a company's performance measurement system. Unlike accounting systems, which must conform to external and internal standards so the figures can be rolled up to final totals, performance measures don't have that constraint. There may be good reasons to enforce some commonality of measures within a process, but in unrelated functions, that should not be an issue.

There are good reasons for decentralizing measurement systems. The first, is that the measures and data systems needed in different processes and sub-processes are so diverse that building a system to accommodate all the needs would be practically impossible. For example, a large financial company which has been involved in quality improvement for several years still has very few performance measures in place because top management wants to have a common measurement system and no one can figure out what that should be. When I asked one of their managers how they can be making any significant progress without performance measures, his reply was, "From what I see, I don't think we are — but we sure are having a lot of meetings."

Another reason for decentralizing measurement systems is that systems customized for different functions are more efficient and more effective than more general solutions. Trying to take the same approach to collecting and processing data in customer service, design, manufacturing, and sales is probably not going to work very well. With the great diversity of performance measures between operating units, it is difficult to see how it would be possible to have centralized measurement systems. Even if it was possible, local control provides greater flexibility, quicker response, ease of use, and more timely reporting than centralized systems.

The best approach appears to be to take a decentralized approach, but to keep closely coupled functions under the same umbrella so the data will share a common structure and can be easily interrelated. Companies that have good measurement systems have generally taken a decentralized approach. Top-level measures that are derived from normal business transactions are obtained from centralized systems, but lower-level measures are left up to individual operating units and departments.

Level of Detail

The amount of detail needed to identify the root causes of problems is typically more than what is required to establish accountability. Theoretically, any process could be measured so extensively that the root cause of any problem could be quickly isolated. Since this could be terribly expensive to accomplish, selecting the right level of detail amounts to making the best tradeoff between how much data to collect and how often the detailed data is needed.

The next best thing to getting to the root cause is to get a specific definition of the problem and enough detail to isolate the cause to a small number of steps in the process. That way, the probable causes can be identified and additional data can be collected to finally determine the real causes of problems.

For example, isolating problems to a particular machine may require 15 process measures, but identifying what is wrong with the machine might require 150 more measures, some of which would be difficult to accomplish. If the machine rarely breaks down, the detail measures are not needed, but if the opposite is true, it would be worthwhile to obtain additional detail measures. Perhaps 10 to 20 more measures would enable isolating problems to different sections of the machine so problems could be more quickly identified.

Focus on Nonconformance or Conformance

Should quality be measured as being 98.5% right or 1.5% wrong? One school of thought is that measuring nonconformance puts emphasis on the negative and will have a negative impact on morale. The contrary view says that 1.5% wrong is very poor quality and that is what should be emphasized.

The consensus of quality improvement practitioners is to focus on non-conformance, because the costs of poor quality are so high. The argument that this will have a negative impact on morale has no merit in my experience and if any papers have been written to the contrary, I have not been able to find them. Since the small portion that is wrong in a process is what messes up everything else, that is what should be emphasized. Can you imagine a hospital bragging that 99.4% of the babies delivered last year weren't dropped on the floor in the delivery room?

Morale always increases when actions are taken that reduce non-conformance. From a motivational viewpoint, the important issue is not whether conformance or nonconformance is measured, but how the performance measures are used.

Combining Measures to Create a Single Composite Index

All managers would like to have one performance measure that would indicate when everything was not in fine shape and tell them what to do about it. It would be nice if that could be done, but it can't. As explained earlier, complex systems cannot be controlled with simple measurement systems. Still, there is at least the merit of convenience in having a composite performance measure for a department or process. Composite measures can also help keep the relative importance of individual measures in perspective by showing that a big swing in a lower-level variable has little impact on the whole process.

The easiest way of constructing a composite performance measure is to assign a weighting factor to each component and calculate the weighted average. Figure 6-3 illustrates developing a composite measure of product quality, based on the customers' perspective of the relative importance of the five different defects. More complicated formulas, such as taking the severity of a problem into account, may also be used.

Composite measures are useful for providing an over-all score, but the following points should always be kept in mind.

1. The level of performance indicated by the composite measure is going to be very dependent on the weights assigned to the individual components. This requires making judgments that may have to be based on limited data. However, if the weights assigned are reasonably correct, the relative importance of individual problems and the degree of improvement in performance are not going to be significantly distorted.

2. As more components are added together, dramatic changes in lower level measures may be masked because the variation in the output will decrease.

If a composite measure is used, it should be accompanied by its component parts. If everyone is not reminded of what makes up the composite measure, it will soon become a meaningless abstraction. Continental Insurance Company experienced this problem when it used a single number to measure performance. When it was introduced, everyone understood its significance, but some employees had trouble remembering how the index was calculated and how it related to their activities.[19] Displaying the individual components also assures deviations in their performance will not be masked by the aggregate measure.

Security of Confidential Information

For performance measures to be effective as a motivator, everyone must be kept abreast of performance — both theirs and the company's. While it is best for performance measures to be available to everyone, it may be necessary to keep some information confidential for competitive reasons. If that is the case, an index relative to last year's performance, an objective, or some other reference will show any change in performance without revealing the actual confidential figures.

Defect	Defect weight	Quantity	Points
A	10	7	70
B	20	3	60
C	5	12	60
D	1	14	14
E	1	10	10
Total		**46**	**214**

Figure 6-3   Calculation of Composite Quality Index. Assuming ten units were produced, the average unit had 4.6 defects and 21.4 quality points. The same number of defects could result in a quality score of 4.6 to 92.0 points, depending on the mix of defects. This table also illustrates that the most frequent problems are not necessarily the most important problems.

SUMMARY

Ideally, implementation of performance measures should start with developing a strategy. However, in practical terms, there are usually some clear opportunities that can be pursued while a strategy is developed. Since few companies have the resources to implement performance measures simultaneously in all functions, selecting pilot areas is a good option. Because implementation is a learning experience, the earlier a company gets started, the earlier it will have a useful system.

Successful implementation requires support from an organization, so management must lay the proper foundation by answering the universal question: "What's in it for me?" After that, determining what to measure can begin. This is also a learning experience, as is developing the data collection and processing systems.

The exact shape any company's system takes will depend on what information systems are already in place, its processes, and what it needs to measure. It is usually better to decentralize measurement systems and tailor reports to each user's specific needs. Measuring the right things is critical, but the ultimate success of a performance measurement system also depends on how easy it is for everyone to furnish the data and use the information provided by the system.

7

ANALYZING AND INTERPRETING MEASURES

A measurement system may be well designed and reliable, but if the data and information are not properly analyzed and interpreted, the benefits provided will be limited. Although rigid rules for analyzing and interpreting performance measures and their related data cannot be defined, heeding the following guidelines will help assure the data is analyzed correctly and the right conclusions are drawn.

"It is a capital mistake to theorize before one has data."

Arthur Conan Doyle

Variation and Statistical Analysis

All performance measures will exhibit some variation. At the lower levels of detail, this variation can be quite large ($\pm$30% or more from week to week), even if a production process is in control. This is called "common cause" variation, because it is an inherent characteristic of the process resulting from the variation in its individual components. In addition, there can be "special cause" variation created by external forces or changes within the process. Not recognizing that some variation is inevitable, will result in managers chasing noise and oscillating between elation and depression when performance has not really changed at all.

The first rule to follow when interpreting performance measures is to not react to short term deviations until the reasons for the deviation are understood. If the deviation is within the normal range, there has been no change in performance at all. If it is a very large aberration, something unusual has happened and the cause should be determined. In most cases, special problems or circumstances will be known by those responsible for the performance measure. Understanding what is happening in one's area of responsibility is one of the first requirements in properly interpreting measures. For example, if it appears performance has improved during the past few weeks, but no changes have been made to processes, the "improvement" is probably a spike that will disappear in the next few weeks.

A simple line graph or run chart will provide a good sense of the normal variation in a process. This is one reason why performance measures should always be put on run charts instead of relying solely on reports. Much of what is perceived by managers as unusual circumstances, is really normal behavior.

> *While discussing the problems of managing inventories, one manager exclaimed: "This business is hard to manage. Take this order, for example. This guy wants 20,000 pounds by Friday but he hasn't ordered this product in six months! We get exceptions like this all the time!"*

The manager was correct in saying that events like the unusual order happened all the time, but that meant they weren't exceptions at all, just normal behavior. This became quite evident when a 4-week moving average showed that demand for each product was actually quite stable as shown by Figure 7-1. Other techniques such as exponential smoothing can be used to reduce the amount of variation in measures and present a more meaningful picture of performance.

Although these simple techniques of dealing with variation are sufficient in many cases, the only way to really understand and properly interpret variation is to use statistical analysis. That subject is beyond the scope of this book, but there are many good books and training courses available on statistical process control. One cautionary note is in order; that is, a little knowledge can be a dangerous thing.

> *My favorite story about the misuse of statistics occurred at a seminar conducted by Edwards Deming. After the seminar, Dr. Deming was answering questions, when a marketing manager approached him. The manager proudly showed Dr. Deming his sales control chart. Dr. Deming took one glance at the chart and gruffly said: "That's a nice chart, but why would anyone in their right mind want a control chart on sales? You don't want sales in control; you want sales out of control — going up! Don't you realize that?!" When last seen, the devastated manager was leaving the room by crawling under the door with considerable room to spare.*

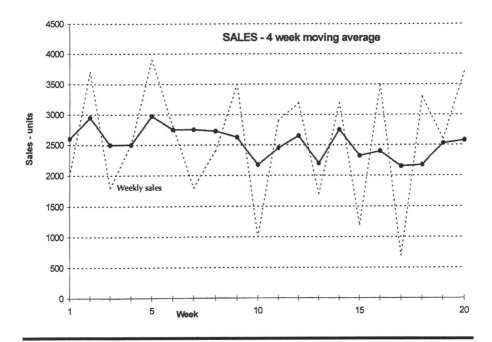

Figure 7-1   Weekly and 4-Week Moving Average of Sales

In a broad sense, if you are using numbers, you are using statistics. Statistical analysis is a powerful tool, but it must be properly used to add value to data. Since improper use of statistics can lead to incorrect conclusions, companies should make sure the people doing statistical analysis have the appropriate knowledge and experience when important decisions or complicated situations are involved. Appendix D provides a good example of how improper statistical analysis can lead to conclusions that have no merit.

Identifying Relationships

Identifying relationships between variables is important for understanding how a process works and also for identifying the causes of problems. Looking for relationships should be part of analyzing any process, especially those that have several external or internal variables that might affect process performance. This also includes customers, because particular customers can influence quality measures such as complaints, returns, warranty claims, and general satisfaction.

Scatter diagrams are a simple way of identifying a relationship between two variables. This consists of plotting points for different values of the

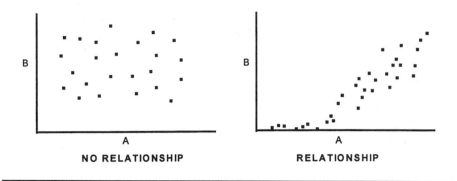

Figure 7-2 Scatter Diagram

two variables as shown by Figure 7-2. If there is a strong relationship between two variables, it will probably be indicated by a scatter diagram. If the relationship is weak, it may not be readily apparent, especially if there are other factors influencing the behavior of the variable of interest.

More sensitive and precise ways of identifying relationships are regression analysis and Taguchi design-of-experiments methods. These techniques can quantify the influence of several related variables on a process or single performance measure. There are several powerful and inexpensive statistical analysis software packages available for using these analytical methods.

Stratifying data is another way of identifying relationships between variables. Stratification consists of cutting the data into layers according to the different variables in a process. For example, to analyze customer complaints about toasters, it might be appropriate to look at them by model number, which plant made them, and which retail chain sold them. In this case, the distribution of complaints to units sold might look like Figure 7-3.

The analysis indicates a possible relationship with model 117B, plant 3, and store chain W, but these relationships are far from certain. It could be chain W just happened to sell many more 117B units than the other models and that plant 3 just happened to make mostly 117B units. It is also possible there is nothing wrong with 117B units, but chain W mistakenly misrepresented the units in its advertisements. Stratifying the data does not supply the answer, but the choice of where to look has been narrowed.

When looking for relationships, it should be kept in mind that any variable in a process has the potential to affect it. Even inputs or process parameters that aren't supposed to change or affect a process should be

Complaints - percent of units sold

Model	%	Plant	%	Chain	%
106A	1.1	1	0.9	K	0.6
117B	2.6	2	0.8	S	0.8
419B	0.8	3	2.4	P	0.7
777A	0.5	4	0.9	T	0.8
				W	2.1

Figure 7-3   Complaint Stratification Analysis

considered. Unless cause-effect relationships are absolutely known, it is not wise to make assumptions before stratifying data. For example, if analyzing manufacturing product quality, stratifying data according to vendors, part numbers, product groups, process step, operator, shift, day of the week, department, supervisor, type of material, temperature in the plant, design engineer, sales representative, customer, or any other associated variable would be appropriate.

Stratifying data can yield some unforeseen results. For example, an analysis of defects in a metal finishing process showed that during one period of four weeks, the defects on one of three plating lines were only one-third of what was normal. Further investigation revealed that during those four weeks, a chemical component that was normally very stable went out of control on the line with the lowest defects. In fact, it was at only half of its specified level. What this meant was that the process ran better when the variable was out of tolerance than when it was within specifications!

Additional investigation revealed the specification may have been correct at one time, but several changes had been made to the process in the past few years. Further analysis and testing determined the original specification was incorrect and the specification should be changed to a figure close to what was discovered by accident. This stratification "fishing expedition" identified a relationship that was totally unexpected and yielded a big return. The problem might have been discovered by other means, but since the specification was not being questioned by anyone, it is more likely the process would have continued running for a long time with the wrong specification before the problem was discovered.

It is also helpful to summarize quality and waste problems by broad categories such as being caused by machines, operators, vendors, customers, external or internal factors, hardware or software, and controllable or uncontrollable. Stratifying data by broader categories is similar to looking at the boulders instead of the rocks. There is no guarantee stratification will reveal anything worthwhile; not looking, however, guarantees any relationships that do exist will not be found.

When looking for relationships, a critical point that must be kept in mind, is that if two variables seem to be related, it does not necessarily mean they are related. If it appears A and B are related, there are three possible reasons for the apparent relationship:

1. A and B are not related at all. The apparent relationship is the result of pure chance or coincidence.
2. A and B are related, but A does not cause B or vice-versa. Instead, they share a common bond through some other variables.
3. A and B have a cause-effect relationship.

In other words, correlation does not mean causality. Just because two things happen at the same time, it doesn't mean they are related. As the story goes:

> *Last Saturday morning, I was aroused out of bed by my neighbor who was in his back yard dancing around, chanting, and beating a drum.*
>
> *"Herb! What in the world are you doing?" I yelled.*
>
> *"I'm doing this ancient African dance to keep the elephants away," he answered.*
>
> *"But there isn't an elephant within 10,000 miles of us!" I replied.*
>
> *"Gosh, I didn't think it would be that effective," says Herb.*

To avoid drawing similar erroneous conclusions, apparent relationships between variables must be verified by understanding how the process works. This may require investigation and even experimentation, but it must be done if the process is going to be understood and improved. Fortunately, most business processes are not as complicated as the causes of cancer, so verifying cause-effect relationships is usually not terribly difficult. Physical processes, however, can be very complicated. Considerable investigation and analysis may be required to determine the relationships between variables in physical processes.

Determining Process Capability

Understanding the capability of a production process is very important for both control and planning purposes. For a process such as cutting sheets of metal to dimensions, statistical methods can be used to determine the capability of the process to meet specifications.[20]

For large and complex processes, the capability question is "What performance level can be maintained by the process?

To answer this question, it may be necessary to select a period that seems to represent normal operating conditions and make a judgment from the performance measures. If managers are staying in touch with what is happening, these judgments can be quite accurate.

In terms of improving performance, determining the capability of a process is often not the question. No matter what it is, it may need to be improved.

Theoretically, when all the special problems in a process have been fixed, the process is operating at its capability. But this limit only applies to the process in its current configuration. In that sense, the real question managers often need to address is: "At what point should attempts to incrementally improve the capability of a process be abandoned in favor of a radical restructuring (or reengineering) of the process?"

Consider a complex piece of equipment, which has ten quality variables that are all not within specification. By bringing each of the variables within specification, quality problems are cut in half. Is this as good as the equipment can do? In its present state, yes. But perhaps an additional improvement could be made to the machine by installing more powerful motors. Is this all that can be done? Not necessarily. Maybe gears could be changed, better tools could be used, or laser-guided alignment mechanisms could be installed. All of these changes would increase the capability of the machine, not just fix something that is broken.

If several changes were made, the performance of the equipment might look like Figure 7-4, which shows the improvement provided by each successive change. It might seem further significant improvements are impossible after making Change 4. This could be true — unless, of course, another way is found to improve the machine's performance.

As this example illustrates, determining when a process has reached its practical limit for incremental improvement, is a matter of judgment.

Event	Defects (%)
Start	10.0
Change 1	5.0
Change 2	4.1
Change 3	3.4
Change 4	2.8

Figure 7-4   Process Improvement Steps

If the person making that judgment understands how well the process is performing and its improvement history, that determination will probably be quite accurate. If no measures are in place, making these decisions is just a guessing game.

Determining Production Capacity

How much work a process can do in a given period is its production capacity, which is a type of capability. Knowing this figure is obviously essential for effective planning and management. When production capacity is exceeded, the following will happen.

1. If the process is limited by equipment capacity, work will pile up in front of the limiting steps of the process.
2. If the process is limited by labor capacity, work may pile up in front of the bottleneck step(s) in the process, but the work may also get done while the quality of the work suffers — rework and rejects will increase.

Consequently, an estimate of production capacity can be derived from quality and production output data. By plotting quality at various levels of production output, as shown by Figure 7-5, an increase in quality problems will be seen when the production capacity of the process begins to be exceeded. In this case, the production capacity is approximately 14,000 units per week. This method is not precise, but it will provide a fairly accurate estimate of capacity. It can also identify bottleneck steps in a process, because their quality performance is going to break down first.

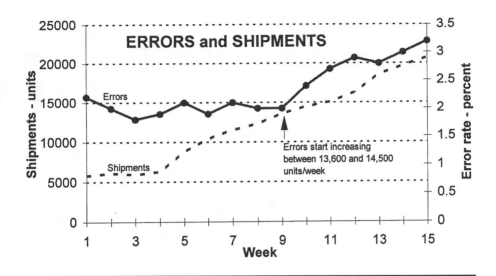

Figure 7-5   Determining Capacity From Quality and Product Output

In situations where workloads are changing because of changes in sales, product mix, staffing, or processes, changes in quality performance can indicate production capacity or capabilities are being exceeded. Although it is usually less pronounced, the same behavior can be observed in equipment performance when it is overworked. When quality problems start to increase and there are no other identified causes, it is a strong indication that the production capacity of the process (or the people in it) has been reached.

Making Comparisons

Executives often want to compare the performance of different departments or units, especially when it is time for raises, promotions, or bonuses. But how can this be accomplished on a meaningful basis?

There are no easy answers. Variance against budget is one common yardstick, but this assumes the budget was valid and the operating assumptions made when the budget was prepared were satisfied during the period in question. The same is true for using actual performance versus objectives as either a measure of performance or for comparing the performance of different units. If the objectives were reasonable and nothing of consequence changed, this comparison is valid, but those conditions are difficult to achieve or verify.

One other basis for comparison is quality. In this case, there are two items of interest — the current level of performance and the rate of change. Comparing rates of improvement is always valid, assuming no substantial changes have been made to what is being measured or how a quality index is calculated. For example, if division A reduced its quality costs ten percent and division B reduced its quality costs fifty percent, it would be a sensible conclusion that B outperformed A in terms of quality improvement. Even if they started at different levels of quality performance, the conclusion is still valid.

Comparing quality performance in absolute terms is another matter. For closely similar operations, valid comparisons are easy to make, but for units that have different products, processes, or both, comparing quality performance is much more an art than it is a science.

In theory, conformance quality of different production processes could be compared by measuring non-conformance and adjusting that figure for the complexity of the process. The common way of measuring non-conformance is in terms of parts-per-million (ppm) which is just a more convenient way of expressing very small percentage figures as whole numbers (0.01% = 100 ppm). This figure can also be converted to a statistical "sigma" equivalent as shown by Figure 7-6. (These figures allow for a 1.5 sigma drift in the process.)

SIGMA	FAILURE RATE - %	PARTS PER MILLION
3	6.6810	66,810
4	0.6210	6,210
5	0.0233	233
6	0.00034	3.4

Figure 7-6   Sigma Rating and Parts-per-Million Equivalent

Statistical methods can be used to determine the non-conformance rate for processes having continuous variables such as drilling holes to given dimensions in a steel plate. For measuring discrete non-conformances such as hitting one's thumb with a hammer, the formula to measure quality in parts-per-million is

$$ppm = (actual\ errors/opportunities\ for\ error) \times (1{,}000{,}000)$$

For example, a process having one step and a failure rate of one unit per thousand produced would have a failure rate of $(1/1,000) \times 1,000,000 = 1,000$ ppm. A process having the same failure rate, but with three serial steps, might have the same 1,000 ppm rating in terms of output quality, but each step would have to be operating at roughly 100 ppm to achieve the 1,000 ppm end result.

Counting actual errors is straightforward, but determining the opportunities for error is often not so easy. For example, how many opportunities for error are there in making a piece of furniture? Do you count the pieces, the individual steps, or what? How about measuring the typing error rate in a report? Do you measure errors per word, per line, per page, or per letter? What you pick is going to make a dramatic difference in the calculated error rate, but the process and its performance characteristics are going to remain unchanged.

An approach that can be used to arrive at a ppm figure, is to assume that before quality improvement efforts start, all processes are operating at four-sigma (6,210 ppm). This is the performance level of the average company and operation as determined by surveys of processes such as writing up orders, airline baggage handling, and payroll processing.[21] Then, the divisor in the ppm equation can be chosen so the initial figure is within the given range.

Consider a complex assembly process that is producing 2,000 units per week, with an error rate of 160 defects per week. On a per-unit-produced basis, this is an error rate of 8% or 80,000 ppm, which seems like poor performance. However, if there are a thousand opportunities for error on each unit, the error rate falls to 0.008% or 80 ppm, which is very good performance. There is no way to determine the actual opportunities for error, but assuming the process is about average, the current defect rate of 8% would equate to 6,210 ppm. This yields the following formula for measuring the assembly process quality performance in ppm:

$$ppm = (defects/units\ produced) \times (77,625)$$

If there were several different assembly plants, they could all use similar formulas, and if they all started improvement efforts at the same time, their relative progress would be directly reflected in their quality ppm index. Everyone will be on the same scale as far as improvement is concerned, but this will not mean that units with the same quality index will necessarily be operating at the same absolute level of performance. Every index will depend on the initial average level of defects (which could easily be off by 10% or more) and the assumption that the starting point represents a certain absolute level of performance.

Comparisons can also be made against benchmarks, but it is not practical to develop good benchmarks for every process. Even if it were possible, unless the benchmarked processes were closely similar, comparing levels of performance would still be a matter of educated judgment. It is possible to make reasonably accurate comparisons against benchmarks, but it requires thoroughly understanding both the company's and the benchmark's processes to be able to make those judgments with any degree of accuracy.

The most valuable comparisons for a company would be comparing its key performance factors to its competitors' and to benchmarks that are truly best in class and/or the best in the industry. This is not easily accomplished, but many companies have been able to do it in selected areas.

Departmental and Process Perspectives

Monitoring performance on a departmental basis is necessary for establishing accountability, providing feedback, seeing appropriate actions are taken, and the development of individuals and the organization. Although this is an important use of performance measures, looking at performance from only the departmental perspective may overlook important process issues.

For this reason, it is important to periodically put all the performance measures of a production process together and review the total process. In this case, the focus should be on identifying the limiting elements and improvement priorities for the total process. When viewed from a total process perspective, priorities may look quite different than they do from a departmental viewpoint.

For example, assume there is a process involving three departments A, B, and C. The weekly quality costs of the top five problems in each department are shown by Figure 7-7A. The top two priorities in each department are relatively clear. However, from a process perspective, five of the top six problems are in A, one is in C, and none are in B as shown by Figure 7-7B. From the process viewpoint, most resources should be allocated to department A, a small portion should go to C, and department B should get whatever can't be used by A and C. Giving each department an equal share of resources would result in applying about one third of them to areas of little return.

Analyzing performance from a process perspective is very important for efficient allocation of resources. However, with the departmental orientation of most companies, it is not likely to happen unless someone is specifically designated to do so.

Department A		Department B		Department C	
Problem	Points	Problem	Points	Problem	Points
A1	1256	B1	457	C1	802
A2	1022	B2	256	C2	421
A3	878	B3	145	C3	303
A4	819	B4	102	C4	254
A5	693	B5	79	C5	102

Figure 7-7A Top Two Departmental Problems Based on Quality Costs

Process	
Problem	Points
A1	1256
A2	1022
A3	878
A4	819
C1	802
A5	693

Figure 7-7B Top Six Process Problems Based on Quality Costs

Zooming In and Out

It is a mathematical certainty that higher level, or aggregate, performance measures will show less variation than their lower level components. The same is true of the time it takes to see changes in performance. For this reason, it is important for managers to periodically zoom in on the detail measures. Probing into trends and large variations in detail performance measures can identify areas where action is needed or where management support would be beneficial. It also increases a manager's understanding of what affects a process and what is happening within it.

Similarly, looking at individual trees is no way to see a forest. Zooming out and taking a broader view of what is happening in the company, department, or process should also be done every few months. In this case, longer-term trends and relationships between measures should be the primary focus. When taking the broader view, questions like the following need to be asked from a total company perspective.

- Where is the workload increasing and why?
- Where has progress stopped?
- Which strategic objectives are being/not being achieved?
- Is any realignment of priorities in order?
- Where do we need to apply more or less resources?
- Which areas look as if they are starting to weaken?
- Which departments or groups need help?

Zooming out also means taking a longer-term view of performance. What may seem to be big gains or losses in performance last month, may look quite different when the past twelve months is considered.

A company that was trying different pricing strategies and promotions to increase customer volume and a product's sales illustrates how different performance can look from a broader and longer perspective. Tests of various offers in a few markets indicated some increased sales 10 to 15%. However, when these promotions were implemented on a broader scale, they initially met expectations but failed to yield any increase in customers or sales over a six-month period. The reason this happened was that the price and promotional changes caused some customers to move their purchases forward in time. This created the illusion of increased customer volume, when actually no new customers were buying the product. The same type of response can be encountered in just about any effort to improve performance. The "Hawthorne effect" of making any change or just giving any performance issue more emphasis will usually result in some improvement in performance. However, if processes weren't changed, the gains will not last.

Zooming out can show patterns that can't be seen by looking at the details. A case in point is a manufacturing production line that was experiencing quality and production downtime problems. Efforts to improve performance were effective in some areas, but the shaping machine seemed to defy any attempts to improve its performance. When performance data for a four-month period was analyzed, it showed repeated problems with gears, chains, motors, and other parts in one of drives inside the machine. Most of the parts had been replaced or adjusted several times in the four-month period.

When this data was reviewed with the operators and maintenance technicians, everybody reached the same conclusion: continuing to repair the drive on a piecemeal basis was not going to work; it was worn out and a major overhaul was needed. The overhaul was accomplished with an all-out weekend effort and performance dramatically improved.

It could be argued that someone should have reached the same conclusion without analyzing any data. Perhaps so, but the repairs made on the machine were spread over three shifts with two maintenance men and one operator on each shift. No one could possibly see the whole problem, and given the level of activity in the plant, no one could be expected to remember the details of what happened weeks ago and put them together. Even with the data, the pattern of behavior was far from obvious because many different parts and actions were involved.

Looking for Opportunities as Well as Problems

The previous example of the plating process, where defects decreased when the chemicals were not within specification, illustrates the value of looking for opportunities as well as problems. While solving problems may be the most important aspect of improving performance, the other side of the coin should not be forgotten. Managers tend to get upset and ask questions when performance is worse than normal, but when performance is better than expected, no one seems to ask why it happened.

Questions like "What is different?" or "What has changed since last week?" need to be asked. The answers may be just as difficult to determine as when trying to solve a problem, but they can also be just as valuable. As with looking for problems, normal variation must be taken into account in order to avoid discovering an opportunity that doesn't exist.

Interpreting Performance Measures

Context

Performance measures do not exist in a vacuum. They are affected by anything that affects a company or its production processes. Weather, strikes, supply line disruptions, unusual customer requests, competitors' actions, and just about anything else on the face of the earth, can cause large deviations in performance measures. Anything of this nature must be part of the context considered when interpreting measures.

For example, if the quality performance of a manufacturing plant has been decreasing slightly for the past six months, it is logical to conclude the manager is doing a poor job of improving quality. But if further investigation reveals that during those four months the plant introduced four major new products, which required major changes to manufacturing processes, his performance may have actually been outstanding. That is why it is good practice to note significant changes in environmental factors or unusual circumstances on charts when they occur. Besides explaining what caused particular behavior, these notes can help managers predict what will happen under similar conditions in the future.

Context is important, but making allowances for poor performance can also be overdone. Minor events should never be used to make excuses for large changes in performance. Furthermore, a logical cause-effect relationship should always be identified when interpreting performance measures. After all, everything can't be blamed on the weather.

The same is true when performance is unusually good. This could also be caused by favorable circumstances. Even if the circumstances are not controllable, understanding what happened can lead to new opportunities. Proper interpretation requires managers to approach the task objectively and resist the urge to attribute all poor performance to uncontrollable factors and all good performance to superior management skills and decision-making.

Two maggots, Fred and Sam, were thrown off a garbage truck as it turned a corner. Fred fell near a dead squirrel that had been thrown under a bush. Sam fell into a storm drain where there was no food at all. After a day-long struggle, Sam managed to crawl out of the drain. Disheveled, skinny, and hungry, he could barely move. As luck would have it, his friend Fred just happened to be passing by. "Hi, Fred! You sure look good!" said Sam. "You're so sleek and fat, but I'm so hungry I could eat nails. To what do you attribute your success?"

"Brains and personality," said Fred. "Just brains and personality."

Quality and Productivity

Quality and productivity are strongly related, but they do not necessarily always move in the same direction. There are four possible combinations of changes in quality and productivity that have different meanings and important implications:

1. **Product quality increases and productivity decreases.** Quality is being improved by brute force, not by improving production processes. Intensive inspection and rework will improve quality, but only at the expense of lower productivity.

2. **Product and process quality increases, but productivity does not increase.** This means performance has improved, freeing up resources that are not being utilized. Putting it another way, quality has improved, but work has expanded to fill the time available. The resources released by improved quality need to be identified and reallocated to other activities.

3. **Product quality, process quality, and productivity are all increasing.** Process improvement is working and resources are either being reallocated or absorbed by increases in the workload. However, this does not mean all free resources available are being used.

4. **Productivity and quality are both decreasing.** To paraphrase a popular beer commercial, it doesn't get any worse than this! The performance measures should tell you where the problems lie. Get them corrected as soon as possible.

Strategy, Operations, and Profits

Strategy, operational performance, and profits are firmly linked together. A simplified summary of this relationship is shown by Figure 7-8. A successful strategy can be viewed as being in the right place at the right time, whether by accident or design. With a strong strategy, it is difficult not to make money, although poor operational performance can still erode margins. On the other hand, a company that finds itself in the wrong place at the wrong time can be operationally excellent and still have very low profits.

Having been associated with companies at both ends of the scale, I can personally attest to the validity of Figure 7-8. Apple Computer is a good example a company going from a strong strategic position to a very weak one, although the quality of the company's operational performance at either point is not clear.

Strategy	Operations	Profits
Strong	Excellent	High
Strong	Poor	Moderate
Weak	Excellent	Moderate
Weak	Poor	Low

Figure 7-8   Strategy, Operations Performance, and Profits

Of course, both strategy and operational performance can have an infinite number of degrees of perfection. Conceptually, this is shown by Figure 7-9. Maximum profit occurs at the point of maximum strength of strategy and operations; minimum profit (or maximum loss) occurs at the opposite extreme. Any point in-between will have some other level of profitability. Although where a company lies on the strategy-operations grid will determine its level of profitability, a given profit level can be attained by multiple combinations of strategy effectiveness and operational performance.

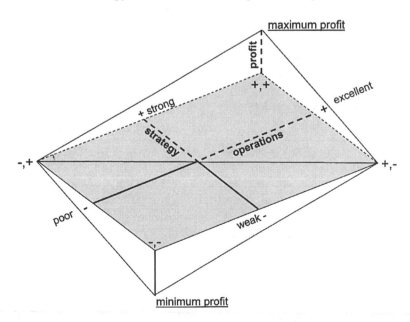

Figure 7-9   The Relationship Between Strategy, Operations, and Profits

Although constructing a strategy-operations-profit chart for a given company may not be practical, the concept is still useful. It says that knowing where a company lies on its strategy-operations grid will determine what type of actions should be taken to increase profits. Excellent operations cannot fully compensate for a weak strategy. Likewise, a strong strategy can compensate for poor operations, but it can also be undermined if it is poorly executed. Since the effectiveness of a strategy is more difficult to assess than operational performance, the quality and productivity of operations should never be an unknown quantity. Furthermore, it certainly makes sense to maintain operational performance at high levels to remove that variable from the performance equation. With proper performance measures in place, even if operational performance deteriorates, it will be a known quantity that can be taken into account when assessing a strategy's effectiveness.

Determining where a company lies on an operations excellence scale cannot necessarily be done in absolute terms, but a good sense of a company's position can be gained from the following information:

- **Company performance measures.** Knowing where you are and where you came from is a prerequisite for determining your place on the strategy-performance grid. Furthermore, knowing how much performance has improved in recent years can be compared to what has happened in related industry groups to get some idea of whether ground is being gained or lost.
- **Information published by industry groups, professional associations, government sources, trade groups, investment firms, and similar sources.** Almost every business sector of any size publishes figures about sales, productivity, and other performance variables that can be used for comparing rates of improvement.
- **Magazine, newspaper, and professional journal articles.** Companies with outstanding operations are frequently discussed in trade and business publications. *Industry Week*, for example, has an annual Best Plants award and devotes an issue of the magazine to that subject.
- **Benchmarking** with companies that have similar processes, but are not competitors.
- **Customer feedback** from third-party surveys and from information collected directly from customers.
- **Competitive intelligence.** A lot of information is available and legally attainable if someone looks and asks for it. It is amazing what can be learned about an industry and companies within it with a little effort.

Xerox, which was in a downward spiral in the early 1980s, illustrates the importance of managers knowing the coordinates of their strategy-operations position. Initially, the company blamed their misfortunes on the Japanese dumping their products on the market. Then they made two shocking discoveries. The first, was that the Japanese were able to sell their products for what it cost Xerox to make them. The second was that the company's cost of quality was 20% of sales revenue.[22]

Clearly, the company's design and manufacturing operations were not competitive. Realizing this, Xerox turned its Rochester, NY plant upside-down and changed all procedures to involve every employee — a move that certainly paid off.[23] Xerox reversed its fortunes and has long since returned to a strong competitive position. The key elements in making this, and other strategic decisions, was knowing its own level of performance and comparing that to its competitors' performance. These initial comparison efforts were later revised to not just look at competitors, but to uncover the best practices wherever they exist.[24]

If a company knows how well it is satisfying its customers' requirements and its performance relative to its competitors', it can make informed decisions about addressing operations, strategy, or both. If it also knows what levels of performance "best-in-class" companies are achieving in key areas, it will know what performance is attainable, how far behind it is, and what it has to accomplish to achieve superiority.

Certainly, all of this information cannot be obtained in precise terms and numbers, but companies have demonstrated that enough quantitative and qualitative information can be obtained to make effective strategic and operational decisions. If a company does not have this information, it is just guessing about where it is and where it needs to go. The performance measures and benchmarks won't improve performance, solve the problems, or define a viable strategy, but they will tell a company where it stands and where performance must improve or other actions must be taken.

The Importance of Market Share

Market share is a key indicator of the effectiveness of a company's strategy and competitive position. Market share can be increased in the short term by decreasing prices, but in the longer term, that is not possible unless a company has a cost advantage relative to the value of its products. The size of a company's market share is important, but in many respects, the most significant aspect of market share is whether it is increasing or decreasing. If it is increasing, it says the company has a competitive advantage in the marketplace. If decreasing, the opposite is true and prompt action is needed to prevent further erosion. Had General Motors, Ford, and Chrysler heeded the signals sent by their small losses of market share in the early 1960s, how different their fortunes might be today! The same is true of the U.S. commercial electronics industry, which has essentially disappeared.

Every company should have measures of market share because it says so much about their competitive position. Getting an accurate measure of market share may be very difficult in many cases. However, a measure of whether market share is increasing or decreasing can be obtained by comparing a company's sales growth with its industry's growth. Changes in market share should also correlate with customer and non-customer satisfaction surveys. If customers are saying they are very satisfied but market share is decreasing, something very important is being missed in the satisfaction surveys, or the surveys are missing a large segment of the market.

Relating Operational and Financial Measures

Operational and financial performance measures should generally track each other, but there are reasons why they could be showing different patterns, especially in the short-term.

- Since there can be significant time lags between changes in operating performance and when these changes appear in accounting figures, operational measures will generally lead, or predict, financial results. Poor quality reported today may not show up as increased costs for several weeks when the actual rework takes place. The effect of returns, extra service calls, and added freight costs may not show up for months. Many of these costs may not be seen at all because they get thrown into broad cost categories.

  Similarly, if productivity and quality increase, it may take months for increased sales to absorb the under-utilized capacity or to reallocate people to other areas. Until then, unit costs may show little or no improvement. In both manufacturing and services, the work-in-process pipeline must be emptied before the full impact of changes in performance will be evident in financial reports. This delay could be as long as several months.
- Financial account structures will probably reflect organizational structure rather than production processes, making it difficult to correlate the two sets of measures.
- Administrative or indirect manufacturing costs may be allocated to goods or services according to formulas that are not related to how processes work. For example, indirect costs are commonly

allocated on the basis of units or dollars produced. This can result in giving low volume products a much smaller share of indirect costs because they often consume a large share of these services.

- Accounting reports aggregate costs that have the same name, even though they may be driven or caused by entirely different events. Freight costs, for instance, may be caused by normal shipments, rush shipments from vendors, or expedited shipments to customers because of problems throughout the production process. Each of these costs is different from the others, but they all wind up in the same bucket of "freight costs." Then, no one is held accountable for these costs, because no one really understands where they came from.
- Since accounting reports are usually produced on a monthly basis, they have a built-in smoothing factor that may mask some changes in performance. Any special charges, adjustments to accounts, or timing problems can also cause financial measures to deviate from operational measures.

Modifying the financial account structure so it coincides as closely as possible to the operational measures, is one step that can be taken to make it easier to relate financial and operational performance measures. It may also be possible to summarize operational performance data in ways that correspond more closely with financial measures. What can be done to better relate financial and operational measures, depends on the systems a company has in place. The primary objective is to verify the accounting and operational measurement systems are tracking each other and if they are not, to understand why. The issue is not which system is right or wrong because they are designed to provide different information and accomplish different objectives.

A secondary objective of relating operational and financial measures is to estimate the financial impact of the key operational measures. Quality costs (the costs of inspection, prevention, rework, and waste) can generally be easily related to financial results. With these costs running 20 to 30% of sales in most companies, quality costs should be a standard component of financial analysis. The impact of customer dissatisfaction on sales and selling costs is not easily quantified, but analysis of sales and customer satisfaction data could yield reasonable estimates.

In order to properly relate operational performance measures with financial measures, managers need to understand both systems as well as how key processes work. This can be a demanding requirement, because the current level of understanding of costs and operations is apparently nothing to brag about in a great many companies. If it were, it would seem the success rate of job-cutting efforts in many companies would be higher than what it has been. Most major corporate downsizings have failed to produce what was expected, according to Peter Scott-Morgan, associate director of Arthur D. Little.[25]

Determining Priorities

Because there will always be more problems and opportunities than there are resources available to pursue them, managers must always think in terms of priorities. Priorities for improving performance, or changes in those priorities, should be one of the regular outputs of analyzing performance measures. Assuming a measurement system has the capability of determining the relative impact of performance variables, priorities should be relatively clear in terms of costs or profit opportunities. However, costs, sales, and profits should not be the only factors considered in establishing priorities. Potential risk, the investment required, the payback period, how well projects support strategic objectives, the availability of required resources, and other factors must also be part of the priority setting process.

From a customer perspective, both the importance of a particular variable and its current level of satisfaction must be considered when setting priorities. As Figure 7-10 shows, where a performance factor lies on an importance vs. satisfaction grid should determine what action is taken. This table should always be considered when determining priorities to prevent resources being allocated to projects that are totally unnecessary in the first place.

IMPORTANCE	SATISFACTION	ACTION REQUIRED
High	High	Maintain current level of performance.
High	Low	Potentially very costly situation. Priority for action and additional resources.
Low	High	Possibly too many resources being applied. Reapply elsewhere, if possible.
Low	Low	Low priority items. Address if time and resources permit.

Figure 7-10   Importance — Satisfaction Priority Grid

Priorities must also be evaluated from the broader perspective of the total company to avoid sub-optimization and to assure resources are allocated to the areas of most return. As discussed earlier, what looks like an important issue from a departmental perspective may be immaterial in terms of the total process. By the same token, where comparatively large returns can be produced with small investments, smaller problems and opportunities should be captured.

A suitable period must be selected when determining priorities. If too short a time frame is selected, the data sample may be too small to be representative; if the time frame is too long, much of the data may be too old to reflect current conditions. In most operating environments, the last 1 to 6 months is usually the most meaningful, but it all depends on the response of the production process and what changes have been made to it. Priorities should be determined from data taken during a period that represents current conditions, normal operation, and is long enough to provide a representative sample of data.

Comparing Results Against a Forecast

If managers understand how a process works, its current situation, and what is being done to improve its performance, they should be able to forecast results with a reasonable degree of accuracy for the next 3 to 6 months. Forecasts will always be too high or too low, but continually missing short-term forecasts by a wide margin indicates a poor understanding of the process and/or the situation when the forecasts were made. When this happens, special efforts should be made by the responsible managers to understand what is causing the difference. Of course, unforeseen events could invalidate the forecast, but exceptions cannot happen all the time.

Look At the Whole Picture

The best safeguard for assuring proper interpretation of performance measures, is to look at the whole picture of performance, not just a piece or two. Businesses and their processes are very complex and their behavior cannot be explained with only one or two variables. The performance of any department or operating unit cannot be judged by the quality or quantity of its outputs alone. The quality of work inputs, vendors, and support services must also be considered, along with the workload and relevant external factors.

Looking at the whole picture, includes looking at longer-term trends in conjunction with most recent performance. Taking a broader and longer view can keep short-term problems in perspective and prevent over-reacting to minor bumps in the road.

A case in point was a vice president who was hyperventilating because he saw a scrap report that listed 200 parts that had been cut to the wrong dimensions. When it was pointed out that the plant produced over 15,000 components that week and the plant manager had drastically reduced scrap and rework during the past year, his attitude quickly changed. Rather sheepishly he said, "Well, I guess it's a good thing I didn't call Bill and chew him out. Maybe I should be congratulating him for the improvements he has made." Probably so, but that should have already happened several times during the past year.

When reviewing performance measures, look at all the measures at the same level to see if they fit together and reflect what has been happening. Mixing lower level detail measures in with key performance measures can lead to confusion and incorrect conclusions. The relative importance of performance measures and their cause-effect relationship must always be kept in mind. It is usually possible to find some detail measure that is exhibiting exceptionally good or bad performance, but this cannot be the explanation for everything happening in a company.

Cause-effect relationships must be understood and performance measures must be reviewed from that perspective to verify the changes in top level measures are explained by lower level measures. If there are any conflicts between the top level and detail measures, then further investigation is required to resolve the discrepancy. Since mistakes can happen anyplace within the data processing chain, anything that doesn't make sense should not be accepted until it is either corrected or explained.

Team Consensus

Looking at the whole picture of performance in terms of breadth, depth, and time will help assure correct interpretation, but there is no way to guarantee it. Given the same information, different people will reach different conclusions because they have different knowledge and experience. For this reason, using a small team of qualified individuals to discuss and reach a consensus on the meaning of performance measures is a good practice. This is best accomplished when the team members do not feel threatened by the performance measures and can feel free to be candid and objective during their deliberations. A culture that encourages honesty, openness, and "telling it like it is," will help ensure performance measures are correctly interpreted.

SUMMARY

Get your facts first, then you can distort them as you please.

Mark Twain

Mark Twain makes a good point, especially for politicians. In business, however, the objective is to get the facts, interpret them correctly, and take appropriate action.

Getting the facts, requires analyzing the performance measures and their data. Moving averages, exponential smoothing, control charts, and other statistical methods are useful for accounting for the normal variation found in performance measures. Scatter diagrams, stratification, and regression analysis are effective ways to identify relationships, but these relationships may be causal or only casual. Any apparent relationship should be regarded as coincidence until a logical connection has been identified or otherwise established.

Other methods of analysis are making comparisons, looking at performance from both a department and process perspective, zooming in and out to see both the trees and the forest, and looking for opportunities as well as problems. The objectives of analyzing performance measures and data are to identify relationships, and to determine process capability, work capacity, and priorities.

After getting the facts, they must be interpreted. Understanding how a process works and the cause-effect relationships between process variables are basic requirements for properly interpreting performance measures. It is also important to understand the relationships between quality and productivity; strategy, operational performance, and profits; and financial and operational performance measures. If these elements are poorly understood, wrong conclusions will be drawn and faulty decisions will be made.

Looking at the whole picture of performance will help assure correct interpretation, but there is no way to guarantee it. Since different people will interpret the same information in different ways, a team approach can increase the likelihood of arriving at the right conclusion. Even if an incorrect decision is made, the performance measures will send signals that what was supposed to happen is not happening, and that changes in strategy or action plans are required.

8

USING PERFORMANCE MEASURES EFFECTIVELY

Performance measures will improve quality, productivity, and profits only if they are properly used. Outside of the initial improvement that normally results when performance measures are first implemented, performance measures will accomplish nothing by themselves. They are only a tool for managers to use.

Like any tool they can be misused and have the potential to do more harm than good. Accordingly, the following guidelines are provided for getting the most value out of a performance measurement system.

> *Even if accurate data are available, they will be meaningless if they are not used correctly. The skill with which a company collects and uses data can make the difference between success and failure.*
>
> **Maasaki Imai**

Verify Accountability is Defined for Each Performance Measure

One of the first steps to take after a measurement system has been implemented is to review the measures to see if accountability is defined for each performance measure. Logically, only individuals who make the decisions or take the actions that affect a performance measure, should be held accountable for its behavior. At the lowest level of detail, each measure should be assigned to a department, as a minimum requirement. Where practical, this should be carried down to sub-departments or even individuals. The only measures which cannot have one, and only one, owner are those in the "other" category where the origin of a performance problem may be unknown.

It is possible that the lines of responsibility for some measures, even for specific quality problems, will initially be unclear because the causes or origins are unknown. Much of this will be eliminated as the process is better understood. Many times, what looks like one problem with several points of origin will later be found to be several different problems.

Even if the causes of problems are known, the lines of responsibility can be fuzzy if processes and tasks overlap work groups. This often happens when processes just evolve without anyone giving much thought to where tasks should be performed. In such cases, it may be desirable to reorganize the work to establish clear lines of accountability. One benefit of implementing performance measures is that it often brings such problems to light.

When reorganizing work is not possible, an alternative is to make the work group with the largest component of a performance measure responsible for all of it. In this case, "accountability" means that the manager is responsible for knowing everything about a process and seeing that its problems are addressed, even if other work groups perform some of the tasks. This situation is relatively rare, but if it exists, it must be addressed. As mentioned earlier, split or joint accountability simply does not work.

In some cases, assigning accountability for a detail performance measure may have to wait until what affects the measure or creates the problem has been determined. In these cases, making an interim assignment is better than just leaving it dangling out in space. However, there is no real harm in letting a few small items be in limbo for a while.

Establish Performance Goals

Short and long term goals should be established for all measures. For key operational and strategic measures, the goals should consider external benchmarks and the time frame to achieve the objectives. Figure 8-1 shows a performance objective for achieving quality parity with a leading competitor in three years.

In the absence of benchmarks or other rational basis for setting goals, there is nothing wrong with being somewhat arbitrary, as long as the goal is reasonable. Where quality is concerned, a useful approach is to set initial short-term goals to reduce defects, waste, late deliveries, or other factors by 50% in six months. Then, when the initial goal is achieved, note how long it took to reach the goal and set a new goal for another 50% reduction in the same amount of time.

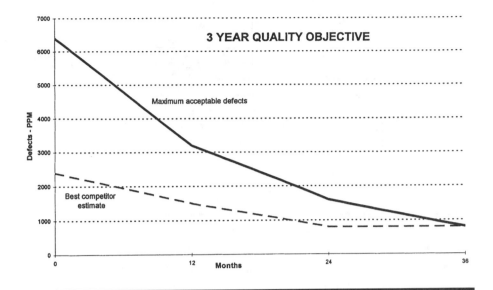

Figure 8-1 Goal for Achieving Parity with Benchmark Company

The reasoning behind this approach is that, in theory, if a company improves a quality measure a certain percentage in a given period, the same level of effort will improve the measure the same percentage each successive period. Some companies that have implemented successful quality improvement programs have demonstrated that this can be accomplished.

Figure 8-2 shows the level of defects that will be achieved each year if the annual reduction rate is 40%. Defects may never get to zero, but after several years they will get so close to zero it won't make any difference.

Goals should be demanding, but achievable. But how do you determine if a goal is achievable? One way to know is if another company has achieved that level of performance (a benchmark). Since benchmarking can take considerable time (years) and effort, goals will have to be somewhat arbitrary in many cases. Companies that have been successful in improving quality, have set ambitious goals. IBM had goals of improving quality performance by a factor of ten in three years and by a factor of one hundred in five years. The Milliken Company set a general corporate objective of improving quality by a factor ten in four years, even after it had won the Baldrige award. In terms of product defects, this amounts to an annual reduction of about 55% per year. What is more significant than setting ambitious goals is meeting them, and goals like these have been achieved many times by companies that really wanted to do it.

Year	Defect rate (%)
1	10.00
2	6.00
3	3.60
4	2.16
5	1.30
6	0.78
7	0.47
8	0.27
9	0.17
10	0.10

Figure 8-2   Effect of Reducing Defects 40% Each Year

Having performance measures without goals is as pointless as driving a car when you have no place to go and you don't care if you get there. That may be acceptable on a pleasant Sunday afternoon, but in rush hour traffic, it just might result in your being run over by some little old lady driving an eighteen-wheeler. The same applies to staying ahead of the competition in business.

Regular Review and Analysis

Since business processes perform tasks and produce outputs on a recurring and regular basis, their associated performance measures should also be reviewed with the same regularity. This does not mean every variable must be reviewed every day, but there should be a reasonably structured review cycle for every variable.

At the front-line level, the review should be frequent (daily, weekly) and address operations performance issues. At higher levels, the reviews can be less frequent (weekly, monthly, quarterly) and should focus more on assessing performance trends, progress toward goals, adjusting priorities, process capability, and strategic issues. Exactly how often any particular measure should be reviewed depends on the response time of the process and the changes being made to it. In the typical business environment, weekly review at the departmental level is generally a minimal requirement, if for no other reason than to keep up with what is happening. With a well-designed system, a manager should be able to identify the important things that need attention in less than 15 minutes.

Regular review is important because it develops continuity of understanding what the measures mean. When interpreting performance measures or deciding what actions should be taken, knowing where you came from and how you got there, is just as important as knowing current performance levels and trends. Performance measures will commonly exhibit both short-term variations and longer-term cycles and trends. While managers should not react to every short-term variation, understanding what caused any recent large deviations can be important to correctly interpreting the measures.

When one of my client companies experienced a dramatic increase in quality problems, an outsider looking at the graphs might have concluded that the company was headed toward disaster. However, knowing most of the problems were caused by tighter quality standards, the introduction of new products, and sudden changes to production schedules, put the numbers in a different light. Obviously these problems must be addressed, but it would have been wrong to conclude there were deeply rooted problems in the process or the organization.

Use Performance Measures For More Than Just Keeping Score

To get any benefit from performance measures they must be used for something other than keeping score. Using performance measures effectively requires more than distributing reports, posting charts on walls, and noting which way the lines are running. "Using" means, among other things, analyzing the information and then:

- Identifying opportunities and problems
- Determining priorities
- Taking action to improve processes and procedures
- Making decisions to re-allocate resources
- Changing or adjusting strategy
- Providing feedback to change behavior
- Recognizing and rewarding accomplishments

Managers should also use performance measures to ask questions about problems, progress, and performance improvement projects to identify where help is needed or what other action might be required. Some questions managers should ask are

- "What were the primary reasons *xxx* was up/down last week/month?"
- "The trend in *xxx* does not show any improvement. What has been done about it and what do you suggest we do to start moving forward?"

- "What do you think are the largest opportunities to improve performance in your department?"
- "It doesn't look like our attempts to improve *xxx* have worked. Have we reached the limits of that process?"
- "What are your top three improvement priorities?"
- "Where do you expect this performance factor to be in three months?"
- "Where do you think we should invest in your department/division and get a good return?"
- "What are the bottlenecks or limiting factors in your department's operations?"
- "What processes are demonstrating a lot of variability in performance?"
- "If these trends continue, what will that mean for costs, delivery, customer satisfaction, and sales?"
- "What can I do to help you with this problem?"

Besides keeping managers better informed, asking relevant questions about operations and performance demonstrates that management understands and cares about what is happening. This dialogue increases the involvement of both managers and employees in managing operations and improving performance. One of the comments frequently made by front line employees about managers is: "They don't have a clue about what's going on around here." That is very often true, because upper-level managers often hear only well-filtered anecdotes about what is happening. Every manager will have a better understanding of what is happening in the front lines if they have good performance measures and use them.

Communicating Performance Information

One ever-present danger with performance measures and their related data is creating information overload by developing and distributing too many reports, charts, and tables to too many people. The quality of information (and its value) is in no way proportional to the volume of information. In fact, the reverse is true.

Data and information are two different things. Data is just raw numbers; information is organized or processed data that a person can use for making a decision. General-purpose reports, which list numerous columns of detail data, should be avoided. Individual managers and supervisors should get the information that is relevant to them, when they need it, and in the form that is most useful for them. What they get must meet their needs and should not be dictated by others who don't understand their problems and processes. Accordingly, individual managers must accept much of the responsibility for seeing that they get the information they need and don't get buried in piles of useless reports.

"Special" reports that may be needed for a short period should not become routine. It is very easy for reports and charts to proliferate to the point where the people using the information become confused. Telling the same story five different ways doesn't make it any easier to understand. For example, if a manager knows the most important issues to address in the department and how those factors have been behaving for the past several weeks, he is well-armed to determine priorities, ask meaningful questions, and start taking action. Reporting each week which variables changed the most from the average of the previous four weeks may be mildly interesting, but what difference is it going to make in what actions a manager should take? Too much information is not information at all — it is just meaningless data.

For effective communication, performance measurement information must be

- **Relevant to the person receiving it.** This requirement has two aspects: Making sure that managers get all the information that is relevant to them and also that they get nothing that is not relevant to them. Information not needed or not used is just another form of waste.
- **Well organized.** Cause-effect relationships, process relationships, and the relative importance of performance factors to the company or operating unit, should be readily apparent.
- **Understood by those using it.** Information that isn't understood is just another form of waste — useless noise.
- **Kept as brief as possible.** Since everyone's time is limited and valuable, the shorter a report is, the more likely it will be used. Wading through pages of numbers to find the important points is not effective use of any manager's time. This is something the data processing system should do by using decision rules and other techniques to separate what's important from what's not.

When processes are not understood, accountability is not defined, and managers don't determine what information they need to do their job, information overload (which is really data overload) results. The obvious cure for this situation is to address these issues and get them corrected.

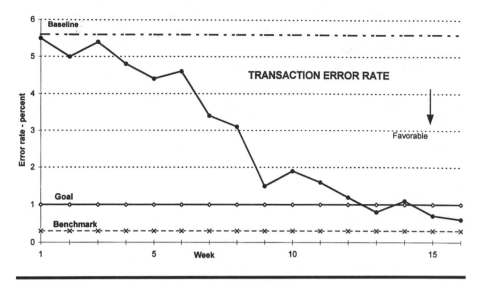

Figure 8-3   Sample Graph

Making Performance Visible

If a picture is worth a thousand words, a graph is worth at least several pages of numbers. Even people who can't read can understand a graph, if they know which direction means better performance. As Figure 8-3 illustrates, a graph enables a person to quickly see current performance, the trend, and the baseline, goal, and benchmark for a particular measure. The graph also lets someone get a good sense of the normal amount of variation in a measure.

The "baseline" is the performance level at some beginning reference point. It is "where you came from," which could be the performance level when a performance improvement program was initiated, at the end of a year, or some other milestone. Sometimes, a baseline can be determined for top-level measures from records that already exist, but normally, it must be derived from new data. Consequently, when first implementing performance measures, enough data should be collected to determine current performance levels before taking any steps to improve performance. Since the "halo effect" of measures will usually cause a rapid improvement in performance and it normally takes a few weeks for data collection to become reliable, determining an unbiased baseline can be difficult. All that can be done is to make the best judgment possible. If the initial baseline is off 10%, it can be adjusted if better information develops. It isn't going to make that much difference in another six months anyway.

The "benchmark" indicates the performance level achieved by a company that represents the best in that particular performance factor. Benchmarks are not absolutely necessary, but since they make a comparison to the outside world, they should be developed for a company's key strategic and operating performance factors. Reducing order time from 60 days to 15 days may seem like a great achievement, but if similar companies have an order cycle time of 3 days, there is still much work to be done.

Graphs should be kept as simple as possible. One variable per sheet is best, but two is acceptable. Any more than that, will be confusing to most people. Run charts will serve most needs, but bar charts, pie charts, and other formats all have their place. When choosing a chart format, first determine what you are trying to communicate.[26] Simple line graphs are good for showing trends, while bar charts are good for making comparisons, such as comparing different periods, products, or regions. Trying to send multiple messages with one chart can be confusing. The graphics capabilities of computers make it possible to produce very sophisticated charts, but elegance should not be confused with utility.

Charts of key variables should be prominently displayed in each department. Since people are most interested in the things they control or which affect them, those are the only measures that should be displayed in their work area. The key measures for the entire division or company should also be displayed in appropriate areas. When charts are displayed, it is important to make any relationships between variables readily apparent, just as would be done in a cause-effect tree. Performance measures of the same relative importance should be on the same level on a display board or their relative importance should be clearly designated. Just sticking charts all over a wall isn't going to get the right message across.

It is very easy for someone who is very involved with a measurement system to unwittingly confuse others who are not so intimately involved. No matter how well conceived and executed, measures that are not understood are worthless. For that reason, a little care in presenting information and explaining what it means, is well worth the effort.

Establishing Priorities

Performance measures will provide important information about a company's priorities, but managers must ultimately decide what to do. Confusion and wasted resources will result if managers don't define priorities and communicate them to everyone affected. There can be both short and long-term priorities, which should be determined using data from the appropriate time frame. Short-term priorities usually consist of addressing current problems, usually "special cause" variations in performance that were seen in the past few weeks. Longer-term priorities should be based on a broader time frame, such as the past several months. Increasing process capability is an appropriate long-term priority.

Reviewing performance should include making appropriate adjustments to priorities and objectives. While priorities should not change very often, a business and its environment are always changing and occasional adjustments will be needed. When priorities are changed, the changes must be communicated throughout a company. This could require changing designated key performance measures and weighting factors in composite measures.

Feedback, Feedback, Feedback

Putting charts on walls and distributing performance reports is necessary, but is not sufficient to change the way people feel about achieving high levels of quality and productivity. This can only come from a company's leaders and managers constantly communicating those values by word and by example. How people act, depends on what they think, but changing a person's values does not happen overnight. There are no definite answers to how long it takes to change an organization's culture, but in my experience, a reasonable time frame for making a significant difference is a minimum of twelve months — and that is with an intensive effort. After that, continuous leadership and reinforcement is still required to prevent backsliding.

To be effective, feedback about performance must be frequent and consistent. Furthermore, much of it has to be on a personal basis. While negative feedback is sometimes appropriate and necessary, it should be the exception rather than the rule. The absolute worse thing any manager can do with performance measures is to use them in a punitive or abusive manner. If that happens, the measures will be seen as a threat, resulting in a swift and subtle reaction. "Accidental" errors in reporting will start occurring which will ultimately destroy the integrity of the entire measurement system.

Most feedback should be positive, especially for accomplishment, but also for trying to improve performance. Positive, encouraging, and meaningful feedback from management will do more to get people involved in improving performance than anything else. Indeed, the most important ingredient in getting results from performance measures, is the leadership and involvement of top management and the immediate department manager.

Reward and Recognition

Asking questions is important, but it is only half of the equation for improving performance. The other half is recognizing and rewarding good performance, improvements in performance, and even efforts made to improve performance that didn't work. Using performance measures as the basis for providing positive, encouraging feedback will do more to boost morale and orient an organization toward improving performance and excellence than anything else. Research has shown that having a person's work and contributions being appreciated consistently ranks among the top three motivating factors for the vast majority of non-management employees.[27]

Again, being specific when recognizing accomplishments, contributions, or efforts has much more impact than generalizations. Although celebrating the accomplishments of a whole company has its place, groups and individuals want to be recognized for what *they* do.

As mentioned before, the big question everyone is going to have about performance measurement and improvement is: "What's in it for me?" If performance improves, employees and managers will expect to be rewarded in some fashion — and they should be rewarded. There are many different ways to link compensation to performance. Discussing the options is beyond the scope of this book, but it must be said that individual piecework performance bonuses are probably the worst approach for promoting continuing improvement and teamwork.[28] Far more effective, are systems that reward longer-term achievements and are group oriented. Rewarding group performance creates peer pressure and promotes teamwork, whereas individual incentives make people want to make themselves look good, even at the expense of others. Whatever approach is taken, quality and productivity values must be built into the reward system in some fashion and everyone must understand this relationship.

Recognition is also a form of reward. It is a more powerful motivator than money, which is generally not considered to be a motivator at all.[29] The best thing about recognition is it costs very little and there are many ways to do it.[30] Performance measures enable managers to give meaningful and relevant feedback to individuals and groups, but managers must make it happen.

How Not To Use Performance Measures

There are some ways performance measures should not be used. Not using performance measures for punishment, humiliation, or confrontation has already been mentioned, but some other uses to avoid are

- **Don't use performance measures to over-control everything that happens.** There is a difference between maintaining adequate control and keeping the lid on so tight that creativity and progress is inhibited. That is not what performance measures are for. They should be a positive force for empowerment, delegation, and taking steps to improve performance, not a feedback mechanism to keep things the way they are.
- **Don't use performance measures to micro-manage.** The ability to see all the details should not be used by managers as a reason to make all the decisions. In the first place, it can't be done. In the second place, it will only alienate everyone who is being micro-managed.
- **Don't use measures only to find something wrong.** The odds are good that out of ten variables, it will always be possible to find some that are temporarily going in the wrong direction. Keep things in perspective and keep a healthy balance between positive and negative comments. It is not a sign of weakness to express satisfaction or delight with good performance. As my mother used to say, "You catch more flies with honey than you do with vinegar."

Performance Measures and Total Quality Management

Readers familiar with the principles and philosophy of Total Quality Management (TQM) will recognize that performance measures will be most effective in that environment. But it is not correct to assume that a company must be deeply involved in TQM to realize large benefits from implementing and using performance measures. To the contrary, any company can benefit from implementing performance measures, but performance measures will move a company toward TQM by:

- Focusing management's attention on satisfying external and internal customers.
- Raising questions about strategy.
- Identifying previously unrecognized quality and waste problems.
- Providing objective information to establish priorities.
- Providing feedback about the success of performance initiatives.
- Getting support from managers and employees for further change when they see tangible improvements in performance.
- Increasing employee involvement by enabling managers to delegate responsibility.

Making Sure the Measures Result in Action

The most basic principle of performance improvement is that if you want to improve the output of a process, you will have to change its input, the process, or both. This requires taking action to change the way work is accomplished, not just looking at the numbers. As some unknown philosopher observed:

"If you keep doing what you've always done,
you'll keep getting what you've always gotten."

One indicator of how well performance measures are initiating action is the rate of improvement in performance indicated by the measures themselves. Another valid measure is the number of improvements made to processes, which is not difficult to monitor. If all performance measures are used for is to keep score, the score will stay the same or get worse.

Although continuous improvement in any performance measure is a desirable goal and theoretically possible, it is also true that the law of diminishing returns, special circumstances, and the efficient allocation of resources may require improvement efforts to be reduced or stopped in certain areas. When improvement efforts have been discontinued, an indicator of how effectively performance measures are being used is the variability of performance. While special causes or situations can also cause measures to go out of control, large deviations are generally a sign that someone is not paying attention to the measures or at least, is not doing what needs to be done. This assumes, of course, that the process was brought under control at some point.

Companies and their environment are in a constant state of flux. A good measurement system will identify when the performance of a process or company is changing because of internal or external factors. As John Mariotti points out so well in *Shape Shifters*, because companies and their environment are constantly changing, companies must change their "shape" in terms of innovation, quality, speed, service, and cost in order to remain competitive.[31] A comprehensive performance measurement system will not tell executives what shape they must take on, but it will signal when changes are necessary.

SUMMARY

Performance measures are only a tool which managers can use properly, improperly, or not use at all. Proper use first requires establishing accountability and objectives so the organization knows where it is headed. Then the measures must be communicated effectively to everyone, reviewed and analyzed on a regular basis, and used in a proactive manner to ask questions, identify problems and opportunities, determine priorities, and take action to improve production processes.

For developing and motivating individuals and groups, performance measures must be used for giving both positive and, in rare cases, negative feedback. Managers must see that appropriate recognition and reward takes place that is consistent with company quality values and objectives. If this does not happen, the benefits realized from performance measures will be minimized.

If performance measures are used properly, quality and productivity will increase and variability in performance will decrease. Although external factors could cause performance problems, when quality or productivity deteriorates, it is a sign that something has broken down in the measurement and management systems. Raising that flag is, of course, one of the primary reasons for implementing performance measures in the first place.

9

INSURING YOUR MEASURES ARE SHOWING AN ACCURATE PICTURE

A perpetual question about any measurement system, including financial measures, is: "How do I know the performance measures are providing me with an accurate picture?" Part of this question can be answered by examining what a performance measurement represents. The rest of the question can't be answered with complete certainty, but by observing the following guidelines, managers can have confidence their performance measures are presenting a reasonably accurate picture.

What Do Performance Measures Represent?

In essence, a performance measurement system is a mathematical model of a company or process that uses data produced by real activities instead of data generated by formulas as in a computer simulation. For a model to be meaningful and useful, it must behave the same way as the system it represents. Therefore, the first question to be answered concerning the accuracy and reliability of performance measures is: "Do the performance measures reflect what we see is happening by other means?" In other words, do the performance measures reflect reality? If they do not appear to be consistent with reality, then the logical conclusion is there is something wrong the measures, not with reality. Of course, it is possible for managers to be out of touch with reality, but hopefully, that will not be a problem.

An example of performance measures not reflecting reality developed in a company where artists interpreted drawings and transferred patterns to films. These films would then be used to make plates for printing colored materials. A system to measure the artists' quality was implemented and after operating for a few months, it seemed to be working quite well. But one day, the department supervisor called to say there was something wrong with the numbers he was seeing in the reports. The measures were saying something was happening that did not correspond to what he was experiencing. Specifically, the performance measures were showing quality had been stable for the past several weeks, but the department was up to its neck in rework.

A check into the measures did not find any data reporting or processing errors, so there had to be other reasons for the discrepancy. A more detailed analysis of the data showed that although the number of quality problems reported during the past two weeks was about the same as reported during previous weeks, the distribution of where they were detected had changed. Instead of being evenly distributed throughout the process, most of the defects reported during the last two weeks were found in the final steps of the process. Since the difficulty of reworking the product increased significantly in the last few steps of the process, the rework workload had increased much more than the quality index indicated. What was needed, was a way to account for the difference in the severity of problems found at different points in the production process.

With input from the department supervisor and the people involved in the process, a weighting scheme was constructed which assigned points to problems depending on the nature of the problem and where it was found. The quality index was then recalculated using the new formula to see if it provided a closer correlation to what the department had experienced. The results are shown by Figure 9-1.

As the graph shows, the quality index derived from the weighted formula differed dramatically from that calculated with the original formula. The new graph showed that quality problems had increased significantly in weeks 14 and 15. Within the limits of human judgment, the chart correlated closely with what the department was experiencing.

A second difference in the charts was that the new formula showed a considerably greater overall improvement in quality than the original. When the department supervisor saw the new graph, he spontaneously exclaimed: "That looks a lot more like what we have experienced. I felt we had actually improved more than the other graph showed because we had put so much emphasis on finding and fixing problems early in the process." Of course, this comment comes from someone who would be biased to think that way, but on the other hand, the supervisor is certainly the most qualified to make that judgment.

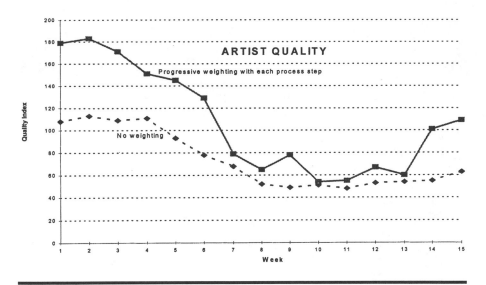

Figure 9-1   Effect of Weighting Formulas on Quality Index. (Top line represents new; lower line, old)

In hindsight, the formula for calculating the quality index should have considered where defects were detected. However, the initial analysis indicated the problems were quite evenly distributed, and neither the supervisors nor the line personnel suggested the cost of rework depended so heavily on where defects were detected. This can be attributed, in part, to the poor understanding of the process before the performance measures were implemented.

When initially asked to describe the production process, the description provided by the supervisors sounded like someone describing a pinball machine. The work was shot into the department where it ricocheted from person to person, depending on what needed to be done to it. After bouncing off enough people, it finally managed to get out the other end of the process. This level of understanding about how production processes work is all too common in companies.

Obviously, applying different weights to the quality problems will change the shape of the graph and show different degrees of improvement. To some extent, it is true that short-term results can be made to look somewhat better than they actually are by manipulating weighting formulas. However, if a weighting formula is constructed from the bottom up, using input from those familiar with the process, the resulting index is probably going to be quite accurate in terms of showing the degree to which performance is getting better or worse. In the longer term, if the formula is far off the mark, the numbers will not correspond to reality at some point.

In this case, it only took a few days to identify the problem and make the adjustments that brought the performance measure into better compliance with what was actually taking place. Fortunately, the supervisor was reviewing the quality results every week, which enabled him to recognize the measures were not reflecting what was actually happening. If he had not been in touch with both the measures and the department's operations, he wouldn't have noticed the discrepancy.

This is a good illustration of the principle that consistently reviewing and using performance measures is the first line of defense in keeping them relevant and assuring they are providing an accurate picture.

Warning Signs

Differences between performance measures and reality can be identified by other means than direct observation. The following warning signs indicate potential problems with a performance measurement system that managers should not ignore.

Abrupt Changes or Discontinuities in the Measures

Large, abrupt changes in the performance of most production processes cannot usually happen without some event occurring which is well known to a number of people. Consequently, when unexplained abrupt changes are seen on performance charts, there is a high probability some error has occurred in reporting or within the data processing system. Any time an unexplained abnormal change in process performance is observed, it should be investigated until it is explained.

A Mismatch Between Measures in a Production Process

When the numbers don't add up or fit together, there is a problem somewhere. For example, it is possible for quality to be increasing while productivity is decreasing, but this is not normal behavior. Another example is if process control measures indicate quality is improving, but the outcome measures of rejects and waste are staying the same. There may be good reasons for this happening, but they must be determined. If something looks strange in the measures, the discrepancy must be resolved. Just saying "That's weird" and moving on, is not appropriate behavior for managers.

Differences Between Financial and Operations Measures

When financial and operations measures are not sending the same message, it should not be assumed that the operational measures are wrong. Accounting reports can be full of timing and reporting errors, even if all accounts balance to the penny. A memorable event I experienced was receiving a monthly accounting report saying a manufacturing plant had a negative waste factor! When the accounting department was advised of the error, the person responsible for the report stood his ground that the report was correct because all the transactions had been processed and the numbers balanced. When he was finally convinced the plant couldn't make its own raw material, it didn't take long for him to find an error in the inventory transactions.

Operational and financial measurement systems provide important checks and balances on each other. My experience is that a good operational performance measurement system is more likely to indicate timing, or other problems, with accounting systems instead of the other way around. However, operational measures can also be incorrect and it is not good practice to jump to any conclusions without first getting the facts. It makes no difference what is causing a discrepancy between financial and operational measures — it must be identified and appropriate action must be taken.

An Increase in the Amount of the Unexplained Performance Gap

An increasing amount of "miscellaneous" or unexplained quality problems is a sign that the process is changing and adjustments need to be made to the measures. Of course, reporting errors could also cause the unexplained gap to increase.

After performance measures have been validated and used for some time, it is easy to assume that all the measures required to measure a process are in place. This assumption should not be made when reviewing performance measures. Variables or important relationships can be overlooked in the measurement system design and variables that initially appear to be insignificant can suddenly become very important. An unexplained deviation in an upper-level performance measure probably means that something is missing from its lower-level measures. If the parts don't add up to the whole, some of the parts must be missing.

Reports and Charts Which are Not Being Used

When people stop using any of a measurement system's reports or charts, it indicates a measurement system is losing its relevance. If someone is not using a report, it means they don't understand it or it is not providing them any useful information. Surveys of users can provide some feedback about what is not being used, but the feedback may be biased because some users may feel that upper management wants to hear only positive answers.

More direct and objective ways of determining what measurement information is being used are as follows:

- Don't issue the reports and see who calls to complain.
- Insert a note on an inside page to call a designated person when the note is found.
- Issue the reports with some pages missing or with obvious errors to see if anyone notices.

These may seem like devious techniques, but they work. It is a good way to cut back on distribution lists, which have an inherent growth rate of their own.

Keeping Up With Changes To Processes

Since a measurement system is a model of a company or other operating unit, any changes to the inputs, processes, or outputs of the unit could require a change in the performance measurement system to keep it relevant. Changes to processes, products, materials, strategy, vendors, customers, customer requirements, or the organization, could mean changes are needed in the performance measures, how the information is reported, or simply how the information is distributed.

As performance improves, changes could also be needed in what gets measured, how variables are measured, how often something should be measured, and how the information is reported. Typically, as larger problems are solved, smaller ones become more important and new problems are discovered. When the boulders and rocks have been cleared away, a person must get closer and use a magnifying glass to see the pebbles and sand. Process improvements can also eliminate the need for measuring some variables. Ideally, all quality problems will eventually be eliminated, making any such measurements unnecessary. It is a comforting thought that in the long run, a performance measurement system should put much of itself out of business.

In today's increasingly competitive environment, businesses must change to survive. As a company changes, its performance measurement system must also change to keep the picture provided by performance measures accurate. These changes do not necessarily have to be major upheavals, but some modifications are likely to be required every year, if not more often. In that respect, the lack of any changes to a performance measurement system over the past 12 to 24 months may indicate the measurement system is not keeping up with the rest of the business.

Improving the Measurement System

Like any business system or process, a performance measurement system should be periodically reviewed to determine where it should be improved. Although a measurement system's effectiveness should be the primary consideration, efficiency should also be addressed. The costs of maintaining and operating a measurement system should not be very significant in the overall scheme of things, but there is no point in collecting and processing data to produce information that is not used.

From an effectiveness perspective, the points to consider are

- Are all key production processes adequately measured?
- Are the proper process measures in place to keep performance within acceptable limits or are only outcomes being measured?
- Is everyone getting all the information they need?
- Are the front line supervisors' and employees' information needs being satisfied?
- Is anyone getting information they do not need?
- If every user could design their own reports, what would they ask for?
- Does everyone thoroughly understand what the reports and charts mean so they can properly interpret the data?
- Are the users taking action as a result of the measures or are the measures just being used to keep score?
- Are the reports and charts in a format that can be easily used?
- Is the information sufficiently timely and accurate for its intended use?
- Can financial results be correlated with operational results?
- Are exceptions or situations needing immediate attention plainly evident in reports?
- When special queries or reports are needed, can they be quickly produced?
- Does everyone have easy access to the information they need?
- Is the relative impact of different performance measures clear to users?
- What changes have been made to processes, products, or quality/service requirements since the last review? Have appropriate changes been made in the measurement system?

In terms of efficiency the following items should be reviewed.

- Is all the data collected actually being used? (Not all data will be used all the time and some data can be collected at essentially zero cost, so less than 100% usage does not necessarily indicate inefficiency.)
- Are there other, easier ways of gathering the necessary data?
- Is there any duplication of collecting or processing data?
- Can the amount of data being collected be reduced by using sampling techniques?
- Can the frequency and/or number of reports be reduced?
- Can some measures be eliminated because they are no longer relevant?
- Are there new data processing tools which can produce the same (or better) information another way?

When looking for ways to improve a performance measurement system, managers should also take a broad look at how the system works. Preparing a system flow chart that identifies all data inputs, processing steps, and outputs can be very helpful in identifying opportunities for improving a system's efficiency and effectiveness.

SUMMARY

After a measurement system has been validated by using it, managers should have confidence in the performance measures. However, it is always wise to be a little skeptical about both operational and financial measures, because significant errors can creep into any measurement system at any point.

If something doesn't seem to correspond with what has been experienced, or if the pieces of the measurement puzzle don't appear to fit together, some investigation is warranted. It may be that one's perceptions of reality are not correct or that the pieces of the puzzle really do fit together when the whole picture is considered. Making that determination is part of understanding the particular production process and learning how to properly interpret its performance measures.

The best way to assure a measurement system is providing an accurate picture is to use it on a regular basis. Staying in touch with both operations and performance measures at the front-line levels of an organization will usually identify when the measures are not consistent with reality. The other warning signs that indicate potential problems with the measurement system, are also more likely to be detected in lower-level measures than in upper-level outcomes. For this reason, performance measurement, analysis, and interpretation must be emphasized at all levels in an organization, not just at the top. If the pieces of a puzzle are accurate, then the picture will be accurate when the pieces are put together.

APPENDIX A

WHAT SOME LEADING COMPANIES ARE MEASURING

The following list of performance measures has been compiled from articles, case studies, consulting engagements, and other long-forgotten sources. They illustrate the variety of quality variables that different companies find important and how they are measured. General categories of measures are first provided and then additional detail is given for each category. In most cases, the performance measures given under each category are not very specific because exactly what would be measured depends on the specific process, product, and customer.

These examples are provided to illustrate application of the process measurement model and to stimulate thought about what should be measured in the reader's business. They are not meant to imply that a company should use any particular measures, or all of them, in its performance measurement system. It is also important to note this is not a complete list or anything close to it. It is only a small sample of what might be measured.

Some of the items listed could apply to more than one category. Where this is true, the measure is listed under what seems like the most appropriate category. However, this may depend on how responsibilities are organized and other circumstances, so there may be some valid differences of opinion about the most appropriate category.

Another point to note is that some of the items measure activities instead of results. Training hours, for example, only mean that people spent time in class, not that anything was learned or that performance improved as a result of the training. While measures of activity can often be meaningless, they do indicate resources consumed and can be leading indicators of performance when that relationship is established. However, the emphasis should always be on measuring results or impact, because that is what counts.

General Categories of Measurements

> Business development
> Customer satisfaction
> Customer service
> Employee development
> Employee satisfaction
> Engineering/design
> Environmental impact
> Flexibility
> Innovation and product development
> Inventory
> Maintenance
> Market share
> Organizational development
> Productivity
> Purchasing
> Quality, external
> Quality, internal
> Sales quality
> Sales productivity
> Schedule effectiveness
> Vendor performance

Note: To make the list easier to read, "percentage of" or "ratio" is implied for all measures, unless otherwise stated. For example, "complaints" means the percentage of transactions that resulted in complaints.

Business Development

- (Also see Innovation and Product Development)
- New business from products, geographic penetration, demographic penetration
- Performance and customer satisfaction ratings compared to benchmarks
- Certifications from customers — percent of business done as certified supplier
- Quality improvement rate

Customer Satisfaction and Dissatisfaction

- Customer expectations vs. company performance — by survey
- Company performance vs. competition — by survey
- Complaints
- Returns and allowances (incidents as well as costs)
- Lost accounts (retention rate)
- Order frequency
- Satisfaction with each aspect of goods or services — sales, technical support, response, quality, value/cost ratio, etc.

Customer Service

- Calls not answered in xx seconds
- Calls on hold longer than xx seconds
- Calls transferred to another party
- Abandon rate — caller gives up
- Inquiry processing time
- Queue time of people waiting in line
- Credit request processing time
- Orders/inquiries not processed within time limits
- Complaints not resolved on first call
- Complaints not resolved in 24 hours
- Degree of satisfaction with complaint resolution
- Courtesy, knowledge, empathy, responsiveness — by survey, response card
- Order entry error rate
- Order fulfillment accuracy
- Backorder rate
- On-time delivery rate
- Orders shipped complete and on time
- Line items shipped complete and on time
- Actual ship date versus requested and promised date
- Actual ship date versus revised request date
- Orders canceled and reason for cancellation

Employee Development

- Employees that have completed a personal development plan
- Employees complying with their development plan

- Training hours per employee per year
- Employees that have improved skills during past year
- Employees certified for skilled job functions or positions
- Employees who have interacted with customers
- Employees involved in planning
- Employees with spending authority
- Employee buy-in to quality improvement
- Employees terminated for performance, other problems
- Increase in average grade level of reading and math skills of employees
- Needs assessment gap — required versus actual skills for positions

Employee Satisfaction

- Attitude surveys to measure satisfaction with many factors — policies, pay, leaders, immediate supervisor, working conditions, training, hours, etc.
- Turnover — voluntary and involuntary, by specific problem
- Absenteeism by reason
- Tardiness
- Employees applying for open positions from particular departments — an indicator of dissatisfaction
- Number of recognition events and awards
- Expenditures on recognition events
- Employees receiving recognition
- Safety measures — accidents, days lost by reason

Engineering/Design

- Design cycle time
- Engineering changes after design completion
- Engineering change orders — by reason
- Improvements to products
- Customer satisfaction with product performance
- Reliability — mean time between failures
- Reduction of parts count on products
- Quality problems attributable to design

Environmental Impact

- Water consumption and/or discharge per product unit, per employee, or per sales dollar
- Waste discharge per product unit, per sales dollar
- Regulatory compliance — audit variances
- Percent of recycled material used as raw material input
- Percent of waste generated recycled
- Energy consumed per unit, BTU/$sales

Flexibility

- Number of standard, common, and unique parts
- Number of different process capabilities
- Percentage of cross-trained personnel
- Production setup/changeover time
- Average lot size being produced — smaller is better

Innovation and Product Development

- Number of improvements made to existing products
- Number of new products introduced/year
- Number of successful new products
- Percentage of sales coming from new/improved products
- Number of new features not duplicated by competitors introduced each year — the number of "firsts"
- Percent of sales from proprietary products
- Patents filed, issued, incorporated into products
- Median patent age in products
- Use of current technology — percent of products made with technology less than xx years old

Inventory

- Service factor — percent of orders filled
- Turns by product and group (aggregate turns is a very gross indicator)
- Production schedule delays because of material shortages
- Inventory items above/below target limits
- Physical inventory variances
- Slow-moving and obsolete inventory

- Excess inventory – anything above normal requirements
- Inventory accuracy and error rates
- Adjustments to inventory records

Maintenance

- Downtime due to different types of equipment failure
- Unplanned versus planned maintenance
- Quality problems due to equipment failure
- Adherence to preventive maintenance schedules
- Waste caused by maintenance tests

Market Share

- Sales/industry sales ratio
- Sales growth rate versus industry growth rate
- New accounts
- Share of key accounts' business

Organizational Development

- Employee and management participation on teams
- Cost reductions, other quality improvements achieved by teams (measured as cost-of-quality savings)
- Employees on self-managing teams
- Employees participating in suggestion plan
- Suggestions/employee
- $ saved by suggestions
- Teams making positive contribution
- Teams achieving goals
- Employees and managers "buying into" quality improvement principles
- $ spent on training as a percent of sales
- Positions filled by internal promotion versus new hires

Productivity

- Sales/employee
- $ produced/employee
- Units/labor hour and labor dollar for direct, indirect, and total labor costs

- Total value of finished products / total production costs
- (Overhead + labor costs) / units produced = value added cost ratio (omits material costs)
- Space productivity — sales or production per square foot

Purchasing

- Quality of vendors — defects by type, returns, delivery performance, etc.
- Percent of parts from certified vendors
- Changes to purchase orders — by reason
- Total number of vendors — lower is generally better, providing quality and cost requirements are satisfied
- Savings relative to previous year costs

Quality, External — What The Customer Sees

- Complaints and compliments
- Credits/returns
- Orders lost — and reason why
- Defects at installation (dead on arrival), during first 90 days
- Defect rate of the total population of equipment at customer sites
- Internal quality — as a leading indicator
- Quality problems detected during product audits in the field
- Percent of bids or proposals accepted
- Technical support costs/unit sold (quality of instructions)
- Mean time between failure
- Service calls or complaints/unit sold
- Revisions to reports for customers — corrections and additions
- Customer quality data
- Certification by customers
- Awards from customers

Quality, Internal — How Well Internal Processes Perform

- Costs of quality — rework, rejects, warranties, returns and allowances, inspection labor and equipment, complaint processing costs
- Waste — all forms: scrap, rejects, under-utilized capacity, idle time, downtime, excess production, etc.
- Yield — net good product produced
- Processes under statistical control with sufficient capability

- Processes "Poka-yoked" or made foolproof
- Process capability (C_{pk})
- Product changes to correct design deficiencies
- Adherence to schedule — tasks being performed on time; jobs ahead or behind schedule
- Changes to purchase and production orders — by reason
- Number of times scheduled ship or complete date changes

Quality Leadership — How Well Executives are Leading Quality Improvement

- Time spent communicating quality values to employees
- Time spent on quality improvement activities
- Accomplishment of quality implementation milestones
- Employee buy-in to quality values and concepts index
- Percent positive feedback from employees after meetings
- Achievement of quality goals
- Quality index based on the Baldrige Award criteria

Sales Quality

- Service, responsiveness, knowledge, empathy — from customer satisfaction survey
- Conformance to company guidelines for lead-times, quantities, special modifications, etc.
- Accuracy and completeness of specifications for orders
- Changes to orders after initial placement — controllable and uncontrollable
- Timeliness and accuracy of price quotations and requests for samples
- Pricing accuracy
- Complaint resolution timeliness and effectiveness
- Response time to inquiries and special requests
- Ethics attribute score

Sales Productivity

- Time spent on selling versus administrative activities
- Sales to selling costs ratio
- Sales process performance — see Appendix C
- Queue/production

Schedule Effectiveness

- Actual versus scheduled complete date by work center, department, production cell
- Late items as percent of average daily production
- Schedule changes — controlled and uncontrolled
- Time lost due to schedule changes or deviations from schedule
- Orders and reports shipped by express services
- Que production time ratio — how long it takes to process an order, divided by how long it would take if there were no delays between process steps (minimum value = 1.0)

Vendor Performance

- Product defects
- Delivery performance
- Quality improvement rate
- Process capability and improvements in capability
- Process quality — from vendors' data
- Cost reduction rate
- Order cycle time
- Emergency response time
- Compliance with operating guidelines, such as having a disaster plan
- Documentation conformance — measured on an on-going basis and by audits
- Billing accuracy
- Time to resolve complaints, get credits for product quality problems

APPENDIX B

IMPLEMENTING MANUFACTURING PERFORMANCE MEASURES — A CASE STUDY*

ABSTRACT

This paper describes the implementation of performance measures in a manufacturing plant to (1) explain the procedures used to determine what to measure, (2) illustrate some of the problems that can be encountered, and (3) compare the costs with the benefits achieved. This case demonstrates that measuring the specific quality problems within a manufacturing process can lead to significant improvements in quality, productivity, waste, and customer service. It also demonstrates that accountability is an important ingredient in an effective measurement system and that performance measures can play a significant role in changing an organization's culture.

The Production Process

The plant produces engraved cylinders, which are used to apply coatings to cardboard and other paper stock during printing. In the trade, they are called anilox rollers. The rollers come in various dimensions, but the most common size is about 2 meters long by 20 centimeters in diameter.

Some rollers are manufactured as an entirely new unit, but most (85 to 90%) are remanufactured. This involves cutting off the old engraved

---

* This paper was presented at the Sixth International Conference on Manufacturing Engineering; November 29 to December 1, 1995; Melbourne, Australia. It is reprinted with permission of The Institution of Engineers, Australia.

surface in a lathe, building up the roller diameter by plating or welding, cutting the roller to the required diameter, and then polishing it to the proper finish for engraving. After the surface has been prepared, it is then engraved using mechanical dies or a laser, depending on the application. After engraving, a chrome, nickel, or ceramic coating may be applied to the surface.

A high degree of precision is required in the product. Typical dimensional tolerances are measured to 0.001 cm. and the engraved surface must be perfect. The smallest defect in the finished surface makes the product unacceptable and it is unlikely any defects can be corrected without repeating the entire process.

In addition to refinishing and engraving the roller surface, remanufacturing may also require repairs to bearing surfaces, journals, or other components, making almost every job unique in some manner. In fact, it isn't until the first several production steps have been completed that what has to be done to rebuild a roller is known. Even then, hidden problems may surface at the worst possible moment.

The plant's production flow is best described as a job shop with a dozen primary routings, ten variations on each primary routing, and another twenty deviations that might occur with any job. With the many decisions and steps required to make a finished product, there is considerable opportunity for both human error and equipment problems.

The plant has 12 work centers and 80 direct labor employees working one primary and two skeleton shifts. Production volume is in the order of 100 units per week with several hundred rollers in production at any given time.

The Organization

The managers and employees would compare favorably with any company. The managers were technically proficient and generally practiced the principles of open communications, delegation, and empowerment. The line employees were quite skilled, experienced, and capable of learning. Except for a few recent immigrants who were learning English, communication was not a problem.

Relations between managers and employees were generally harmonious and cooperative in a non-union environment. An employee survey, conducted shortly before the program started, indicated a generally positive relationship between managers and line employees. However, notable dissatisfaction was expressed with regard to:

- A lack of concern about quality — employees accepting poor quality work and managers not setting higher standards (a point the CEO was clearly taking action to correct)
- Not everyone being held accountable for the quality of their work
- Departments not working as a team and department managers being too protective of their territory

My initial impression of the organization is one which I believe applies to most companies — everyone was trying to do a good job, but frustrated by having to deal with problems and crises they knew should not be happening. It was also apparent that with respect to production problems, the organization was predominantly in a fire-fighting mode, treating each problem as an isolated occurrence. There appeared to be little under-standing of the exact nature of the problems, their causes, and their relative importance.

Both the CEO and plant manager felt measurement of the plant's operations would be necessary to improve performance, but they did not have a good idea about what to measure or how to measure it. A big concern was whether it would be possible to get reliable data from the employees on the shop floor.

Conducting quality improvement training enhanced the positive factors in the plant's culture. Every manager and employee in the company participated in order to lay the foundation for implementing performance measures and getting started on continuous quality improvement. The objective of the training was not to create experts, but to give everyone an understanding of why improving quality and productivity was impor-tant, the basic principles of quality improvement, and the role performance measures would play in the process.

In my experience, the initial reaction to the program was about what can be expected in most change situations — some people were enthused, some neutral, and perhaps a few were negative. It is estimated that 70% of the employees, managers, and supervisors were quite positive about the concept of performance measurement. Most of the remaining 30% would have said: "I'll go along with it, but I'm not convinced it will work or that it is a good deal for me." There was some apprehension about the measures being used to punish people. A few individuals may have been very opposed to the program, but this was never expressed.

This positive attitude toward the program did not just happen. The CEO spent considerable effort explaining the need for the company to improve performance, the importance of having reliable information, and how everyone would benefit from improved quality and productivity. The threatening business climate of "downsizing" that existed at the time also gave everyone a clear reason to support the program. When survival is at stake, change is certainly more readily accepted.

Although the overall climate was favorable, when implementation started, virtually everyone was wondering if performance could be measured and if it would actually improve the situation.

Determining What To Measure

Key Performance Factors

Preliminary studies of the plant's operations and interviews with the supervisors indicated the primary opportunity for improving performance was in reducing internally generated product defects. This might seem to be an obvious conclusion, but in-process product quality is not always the most important issue affecting manufacturing performance. For example, a plant might have very few internal quality problems, but could still be performing poorly because of problems created by product design, purchasing, inventory control, or other support functions.

On-time delivery and reducing order cycle time were also recognized as significant areas for improvement. However, reducing internal rejects and rework had to come first because direct efforts to improve delivery or cycle time were not likely to produce good results until manufacturing operations had been brought under better control. Besides, improving internal process and product quality would certainly have a positive impact on deliveries and order cycle time, as well as costs.

Therefore, the key performance factors for the plant were determined to be

- Internal product quality - rejects and rework
- Productivity and labor costs
- On-time deliveries
- Order cycle time

A general model for measuring the performance of any production process is given by Figure B-1. Relative to the model, the specific items to be measured for the rebuilding process are given in the following table.

Although the top-level performance measures are most important from a results viewpoint, they only provide "score-keeping" or symptomatic

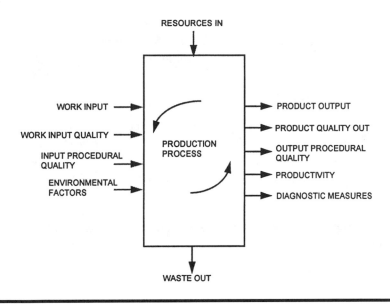

Figure B-1   Production Process Measurement Model

Variable	Performance Measure
Work input	Square inches to be shipped each week
Work input quality	Damaged rollers received
Input procedural quality	Incomplete or incorrect instructions from sales
Product output	Sq. in. shipped each week
Output product quality	Defects found at final inspection and customer returns
Output procedural quality	On-time delivery and order cycle time
Waste	Utilization of critical equipment (material waste was not a significant factor)
Productivity	Sq. in. produced per direct labor hour
Environmental factors	None
Diagnostic measures	In-process defects — by specific problem and responsible operation

information. They indicate the size of the general problem and whether it's getting better or worse, but provide no information about what is causing the problem. That is why lower-level diagnostic measures are needed.

Diagnostic Measures

The key to success in improving performance is to have diagnostic measures, which enable problems to be isolated to the smallest area possible. In manufacturing, this normally means at least getting down to the level of the individual operation — and it may be necessary to get to a lower level of detail such as the individual machine, operator, or product.

Identifying the in-process defects to measure, started with asking the department supervisors to develop a list of quality problems they had experienced in recent months — a "quality problem" being anything that caused rework, rejects, or interfered with a task being performed in an optimal manner.

In essence, the supervisors were told to consider themselves "customers" and identify anything objectionable they were receiving from their "suppliers," whether it was a product problem such as bad copper plating, or a process problem such as defective tooling or incomplete instructions.

When the supervisors completed their lists, they were merged into one master list. Each problem was then assigned a code number according to the department that caused the problem. Next, forms for recording the problems encountered during the workday were prepared for each department. These forms reflected the known problems individuals could experience and provided space for reporting problems not already identified (see Figure B-2).

The forms were given to everyone in the plant along with instructions to list *anything* about the product or process that was not the way the person thought it should be.

As the forms were turned in each day, the supervisors reviewed them to identify reporting omissions, errors, and new problems encountered. This review could only be done by the supervisors, because they understood the plant's processes and the terms used in the plant to describe problems. Someone not intimately familiar with the various operations would not have been able to properly interpret the notations made by the operators.

As additional problems were identified, the forms were revised to include them. After a few weeks use, the list of defects stopped growing, indicating that nearly all recurring problems had been identified.

COPPER PLATING DEPT QUALITY REPORT

WEEK BEGINNING ___/___/___   NAME _____

Description	Notes	Code	Count
Bad turning lines		101	///// ////
Chatter marks		102	///
Cut more than 0.20		103	///// /
Damaged base		104	//
Damaged journals		105	////
Groove*			///
Turn lines in base*			//
	*Operators would add any defects not listed		

Figure B-2   Initial Department Quality Log

System Implementation and Development

Data Reporting

The first obstacle to be overcome was getting reasonably accurate data from the operators. This required them to recognize a problem, interpret it correctly, select the proper problem code number, and enter it on the quality log. In many respects, this amounted to creating a new language in the plant because everyone had to learn how to identify particular problems and call them by the same name. For most employees, this was not difficult, but it took several weeks for some to learn how to report quality problems correctly.

Having checkpoints in the data entry process and feeding back corrective information to the people involved, quickly improved the quality of reporting. The critical steps in this process were

1. Before sending the production logs to data entry, they were reviewed by the responsible supervisor to verify the necessary data had been entered correctly. In most cases, the supervisor could verify the proper code had been entered, but this was not always possible.

2. When the data was entered into the computer, the program would verify the defect code was valid. If not, an error message would be printed and the data log would be returned to the supervisor for correction. Then, the correct code would be determined and the person who had made the error would receive appropriate instruction. This feedback would take place every day.

3. When required, changes were made to the defect list to give a problem a more easily understood name or to add newly identified problems. Additions were also required to break general problems down into their more specific components. For example, what was originally called "stains" turned out to be different problems that occurred at several points in the process, with each one having a different cause. For these reasons, the initial list of 70 defects grew to about 130 in several weeks and eventually reached 163 items.

4. The quality reporting form was revised several times to make it easier to use. It was finally combined with an existing production log, resulting in the form shown by Figure B-3. This eliminated one piece of paper and worked quite well.

The list of possible defects shown by Figure B-4 is printed on the back of the production log for reference. They are grouped and coded according to the department or operation responsible for the problem. Except for damage, the nature of the defect would identify its source, even if the problem was not discovered until a few steps later in the process.

It took about four weeks to get reporting to the point where the data was quite reliable and the defect list was essentially complete. At this point, confidence in the information provided by the system rapidly increased. Of course, the data would never be 100% accurate, but a few errors and omissions would not materially affect the usefulness or integrity of the system.

ANILOX PREP PRODUCTION REPORT DEPT _MILL ENGRAVING_ NAME _BUTCH MESWICK_
WEEK BEGINNIG _10/12/94_

DATE	CONTROL NUMBER	FACE + LENGTH -	MAX/MIN DIAM.	DIAM BEFORE PROCESS	TURNED GROUND POLISHED	HIGH/LOW DIA AFTER PROCESS	MACH NO	SQ IN	N-new D-redo N D	DEF 1	DEF 2	W-rework J-reject W J	BASE CODE	START FINISH
12	82344						506	1236	N				COP	
12	82563						510	2042	D	107		J	COP	
12	80726						512	1152	N	143	145	W	MIL	
12	84536						511	963	N				MON	

NOTE: Only data fields relevant to quality performance measurement are shown filled in.

Figure B-3 Production and Quality Log

ORDER ENTRY/SALES

001	Instructions incorrect
063	Damaged B/S
064	B/S undersized
065	Damaged threads
099	Sales misc.

PREP LATHE

101	Bad turning lines
103	Damaged base
107	Glue/ink on roll ends
109	TIR - prep lathe
110	Turning lines in base
111	Turned too close to min
112	Groove
113	Header B/S undersized
143	Damaged B/S
145	Damaged threads
147	Damaged face
149	Damaged journal
199	Prep lathe misc

GRIND & POLISH

152	Chatter marks
154	Polished thru
155	TIR - Grind & polish
157	Diamond grind - chatter
158	Diamond grind - taper
160	Polishing nicks
161	Polishing undersized
162	Taper
163	Damaged B/S
165	Damaged threads
167	Damaged face
169	Damaged journal
198	Grind & polish misc

COPPER BUILDUP

201	Hole/sink/crack
202	Loose copper
203	Underplated
204	Stopped in plating tank
205	Bad copper
206	Etched headers
207	Nickel showing thru
208	Stains from nickel
209	Flaking nickel
210	Damaged in nickel flash
211	Oversized copper
212	Blister
263	Damaged B/S

265	Damaged threads
267	Damaged face
269	Damaged journal
299	Copper buildup misc

COATING

301	Monel spray loose
302	Undersized - monel
303	Oversized - monel
304	Holes - monel
305	Bad monel
321	Stainless spray loose
322	Undersized - stainless
323	Oversized - stainless
324	Holes - stainless
325	Wrong weld wire
342	Bad ends from welding
343	Undersized - ceramic
344	Oversized - ceramic
345	Holes - ceramic
348	Not sealed
349	Bad center
350	Porosity
351	Taper
352	Holes - weld
353	Rough ceramic
354	Knots
355	Lines in ceramic
356	Ceramic on headers
357	Ceramic on journals
360	Balance holes open/leak
363	Damaged B/S
365	Damaged threads
367	Damaged face
369	Damaged journal
370	Stain before sandblast
371	Stain after sandblast
372	Stain during coating
373	Stain after coating
375	Stain before engraving
376	Stain after engraving
399	Coating misc

CERAMIC POLISH

406	Taper - ceramic polish
463	Damaged B/S
465	Damaged threads
467	Damaged face
469	Damaged journal
470	Stain before sandblast
499	Ceramic polish misc

CHROME PLATE

502	Rough chrome
503	Chrome flaking
504	Chrome missing
505	Chrome bubbles
563	Damaged B/S
565	Damaged threads
567	Damaged face
569	Damaged journal
599	Chrome plate misc

MACHINE SHOP

601	No prints for fab rolls
602	Turning lines - journals
663	Damaged B/S
665	Damaged threads
667	Damaged face
669	Damaged journal
699	Machine shop misc

MILL ENGRAVING

701	TIR - mill engraving
703	Spirals
704	Damaged face-mill engr
705	Mill marks
706	Rust
707	Excessive oil
708	Stains
710	Blew up
711	Bad start
714	Operator misc
763	Damaged B/S
765	Damaged threads
767	Damaged face
769	Damaged journal
799	Mill engraving misc

CAUSE UNKNOWN/MISC

908	TIR - origin unknown

Figure B-4   Anilox Defect/Quality Codes (All codes not shown)

At first glance, it might seem unrealistic to expect front-line employees to report each of 163 possible defects correctly. However, any one person would see only a small number of all the possible problems — usually less than dozen. For this reason, after a few weeks, nearly everyone had memorized the defect codes they would normally use and reporting errors rapidly diminished.

Distinguishing Rejects From Rework

As soon as reporting had become relatively routine, another level of detail was added to the system. This was to differentiate between defects that caused rejects and those that only caused additional rework of the product. Since there was a big difference in cost between rejects and rework, this difference had to be captured by the measurement system.

When reporting of rework was added to the system, another problem was created — half of the jobs in the plant were being reported as requiring rework! The problem was that many of the steps in the process required minor finishing touches that were being reported as rework.

What was needed, was an operational definition that would distinguish normal finishing touches from significant additional effort. After considering other approaches, the rule was established that any problem requiring more than 15 minutes of additional work would be reported as rework. This definition solved the problem.

Multiple Defects

Another problem encountered, was how to report product defects when more than one was involved. How should you account for a defective roller that has two defects, either one of which is sufficient cause for rejection or neither of which, would be cause for rejection by itself, but when combined, create an unacceptable product?

To accommodate this situation, the system was changed to accept two defect entries for the same incident and count them as separate items. This could slightly overstate the number of defects, but since multiple defects rarely occurred and most of them involved two serious defects, the potential error was negligible.

Accountability

Another significant change made to the system related to reporting damaged rollers. After the system had been in use about six months, analysis of the data showed that one of the major causes of rejects and rework was damage due to handling or accidents. Since the damage was happening throughout the plant, it appeared that carelessness was a major factor. Attempts to reduce this damage by bringing it to everyone's attention had no effect, so another approach was needed.

One of the suspected weak points in the system was that damage defects were grouped under the "Miscellaneous" category, meaning no one was truly accountable for them. To establish accountability, the damage defects were eliminated from the "Miscellaneous" group and given a unique code number under each department (see Figure B-4).

A rule was also established by the supervisors that the department receiving the roller was responsible for checking it and reporting any damage when it was received. If damaged, it would be assumed the damage was caused by the previous operation and reported as such. In effect, the "customer" department became an inspector with a vested interest in reporting any damage. If not reported by the receiving department, the damage would probably be reported by the next operation and be assigned to the department that should have reported it in the first place.

Although there were doubts this approach would work, damage defects decreased 70% in two weeks!

This episode is a good example of the power of performance measures when they are combined with accountability. When people know what is important, that they are accountable, and they can do something to affect the outcome, performance always improves. "What gets measured gets done" is often heard in management circles, but I would add, "...only if someone is clearly accountable for what is being measured."

It is important to note that "being held accountable" did not mean anyone was punished for causing damage. What it did mean, was that damage defects counted in the department's total score, which was posted on the plant's bulletin board.

No doubt peer pressure and simply being more careful played a role in reducing damage, but so did several changes made to handling and storage procedures which resulted from knowing where the damage was occurring.

System Output

The primary report provided by the system summarizes defects by responsible department and ranks them individually by frequency. This report (Figure B-5) is generated weekly to provide a timely recap of performance. It is available early Monday morning and enables problems to be discussed while they are still fresh in everyone's mind.

ANILOX DEFECTS BY OPERATION - 10/1/94 to 10/31/94

Code	Group/Operation	Qty	Pct
000	Sales	5	6.4
100	Prep lathe	5	6.4
150	Grind & polish	8	10.3
200	Copper buildup	6	7.7
300	Coating	5	6.4
400	Ceramic polish	1	1.3
500	Chrome plate	9	11.5
600	Machine shop	0	0.0
700	Mill engraving	27	34.6
800	Laser engraving	9	11.5
900	Unknown/misc.	3	3.8
	Total	78	100.0

ANILOX DEFECTS RANKED BY FREQUENCY - 10/1/94 to 10/31/94

Code	Qty	Pct	Description
710	12	15.4	Blew up
799	7	9.0	Mill engraving misc.
503	7	9.0	Chrome flaking
801	5	6.4	Machine failure-laser
802	4	5.1	Wrong engraving
99	4	5.1	Sales misc.
908	3	3.8	TIR-origin unknown
711	3	3.8	Bad start
167	3	3.8	Damaged face
155	3	3.8	TIR - grind & polish
109	3	3.8	TIR - prep lathe
703	2	2.6	Spirals
350	2	2.6	Porosity
209	2	2.6	Flaking nickel
208	2	2.6	Stains from nickel
714	1	1.3	Operator misc.
713	1	1.3	Machine problem - ME
704	1	1.3	Damaged face - ME
567	1	1.3	Damaged face - CP
499	1	1.3	Ceramic polish misc.
352	1	1.3	Holes - weld
345	1	1.3	Holes - ceramic

Figure B-5   Defects by Group and Frequency Report

The ranking of the specific problems is the most important system output, since this provides a clear picture of where action is needed most. The report can be produced for any selected period such as the past 30, 60 or 90 days. Getting a longer-term perspective on problems is essential for setting priorities.

Reject and rework detail is provided by the report shown in Figure B-6. It summarizes all quality problems caused by a department for a selected period and provides both unit and square-inch data to account for size variation. This report is generated monthly and on demand.

The weekly figures for production, rejects, and rework (all measured in square inches) are used to produce graphs of quality (defects as a

COPPER BUILDUP DEFECTS - 10/01/94 to 10/31/94

Defect	Rej	Description	Base	Rptdate	Cntrl	Mach	Sqln
201	W	Holes in copper	COP	10/3/94	81791	117	963
208	J	Stains from nickel	COP	10/11/94	82344	124	1236
208	J	Stains from nickel	COP	10/11/94	82348	117	1236
209	J	Flaking nickel	COP	10/25/94	82675	123	2042
209	J	Flaking nickel	COP	10/26/94	82893	117	2042
210	J	Damaged in nickel flash	COP	10/11/94	81791	125	963

UNITS REJECTED	SQ. IN.
5	7,519
UNITS REWORKED	SQ. IN.
1	963

COPPER BUILDUP DEFECTS - 10/01/94 to 10/31/94

Defect	Rej	Description	Base	Rptdate	Cntrl	Mach	Sqln
350	J	Porosity	CER	10/07/94	81818	411	1152
350	J	Porosity	SPR	10/18/94	82671	410	1009
321	J	Stainless spray loose	SPR	10/20/94	82636	156	253
345	J	Holes - ceramic	CER	10/20/94	82453	411	498
352	J	Holes - weld	SPR	10/20/94	82285	708	1014
303	W	Oversized - monel	MON	10/27/94	82165	412	963

UNITS REJECTED	SQ. IN.
5	3,926
UNITS REWORKED	SQ. IN.
1	963

Figure B-6   Reject and Rework Detail Report

percentage of production) and productivity (square inches/labor hour). The same figures are produced for the individual departments.

Other data and measures used are given below.

Data/Measure	Source
Production & Shipments	Production control system (sq. in.)
Labor hours	Accounting system
Productivity	Calculated from above
Equipment utilization	Run time meters and operation logs

The database system used to generate the reports makes it relatively easy to analyze defects by type, size, roller base material, machine number, or other product or process parameters which are in either the quality measurement or production control systems. This type of analysis is performed when investigating particular problems and has identified relationships that have been helpful in identifying the root cause of some defects.

The measurement system enables managers to quickly review the performance of the plant and each department for the previous week or any period. However, the factors that were most important in improving the plant's performance are

1. Being able to determine the magnitude of the general problem and the relative importance of specific problems so resources could be allocated to where they would provide the best return.
2. Being able to measure the impact of changes to procedures, materials, or other aspects of production processes.
3. Establishing accountability for specific problems and providing timely relevant feedback on performance to everyone concerned — especially the front-line employees and supervisors.

Impact On Performance

Benefits

Figure B-7 shows the decline in total defects (rejects + rework) as an index relative to units produced. From October 1993 to July 1995, internal defects have conservatively decreased 35%.

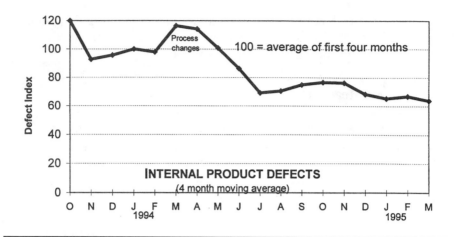

Figure B-7   Improvement in Internal Product Quality

As the graph shows, defects decreased about 20% in the first two months. This is a typical response to introducing performance measures and is the result of: (1) communicating to everyone what is important by measuring it, (2) identifying what needs attention most, (3) establishing accountability, and (4) providing feedback about performance.

The rise in defects in March and April 1994 was caused by changes in processes and abnormal problems with a few pieces of equipment. Since August 1994, little progress has been made due to the plant being overloaded with work because of increased sales. Theoretically, another 35% decrease in defects should be achieved from October 1994 to October 1995, which would put the total reduction in defects at roughly 60%. When the improvement effort is revitalized, there is little doubt additional gains will be achieved.

The reduction in defects is reflected in a 15% increase in direct labor productivity and an increase in the utilization of critical machines from 70% to 85% — primarily because job setups have been reduced and production scheduling has become more reliable.

Improved performance also shows up as a reduction of order cycle time from six weeks to four. On-time delivery performance increased from 75% to 90%, although some of this can be attributed to a better production tracking and planning system.

In terms of labor alone, production costs have been reduced about $350,000/year. This, however, is not the whole story, because improved delivery performance and reduced order cycle time has had a positive impact on sales. How much can only be guessed, but management estimates sales have increased 10% due to these factors.

Costs

Detailed cost records of implementing the performance measurement system were not maintained because developing and implementing the system was intertwined with other activities. The estimated cost is 30,000 to $40,000.

With fewer than 10 quality transactions per day to be entered into the database, operating the system is quite inexpensive. The total annual cost to review and enter data, print reports, produce graphs, and conduct special analyses is less than $15,000. As shown by the table below, the system is a real bargain by any measure.

Measure	Before	After	Change
Defects	100	65	(35%)
Productivity	100	115%	+15%
Machine utilization	70%	85%	+15%
Order cycle time	6 weeks	4 weeks	(33%)
On-time delivery	75%	90%	+15%
Sales			+10%
Direct labor costs			($350K/yr.)

Could the improvements have happened without the performance measures? No doubt some improvement could have been made without them. However, since the company had not been making progress and the problems were apparently not well understood before the measures were implemented, it is difficult to imagine significant results being achieved without them.

Impact On the Organization

To a casual observer, few differences would be apparent in comparing the "before" and "after" pictures of the organization. But a closer look would reveal some fundamental differences in the thinking of the managers, supervisors, and most of the line employees.

Fire-Fighting Reduced

Perhaps the most important change is that the organization is no longer in a reactive mode to problems. The emphasis has switched from "Fix it and ship it" to "Find the problem and solve it." Of course, some problems occur which must be corrected by whatever means possible, but weekly meetings are now held to review performance and set priorities, using factual information instead of just someone's opinion.

Managers say they now have a much better understanding of problems in the plant, especially in the areas where manual skills are important. I believe most managers, particularly upper-level managers, would make this comment after a few months experience with good performance measures. They provide insight into operations that cannot be obtained from other sources.

Increasing Manager's Willingness to Listen

Perhaps the most interesting comment about the project was made by the CEO — that before performance measures were implemented, managers were not as receptive to employee input and ideas as they are now; the reason being that the measures provided an objective and more complete picture of what was happening in the plant.

When front-line employees complain about the problems they must overcome to do their job, their comments are often taken by their superiors to be excuses. Furthermore, the people causing the problems can — and most certainly will — dismiss them as isolated incidents.

But when objective measures are in place, the scope and frequency of problems quickly becomes apparent. Good measurement systems bring to light problems that otherwise aren't seen or are blamed on the wrong person. For example, I have seen a plant manager blamed for missing production schedules when Purchasing didn't supply the necessary material, the schedules provided by Production Planning were hopelessly out of sequence, and the drawings were full of errors.

It is not an exaggeration to say that without comprehensive performance measures, managers really don't know what is happening in their areas of responsibility. As the CEO said, "When we started getting the numbers, we realized most of the problems we heard about were both real and important."

Morale

Although no studies were conducted, it is safe to say the organization's morale has definitely improved. The most probable reasons for this are the sense of accomplishment gained from improving performance, a reduction in the stress caused by quality problems, creating common goals, and getting everyone to focus on problems instead of personalities.

Another aggravating factor that has been removed is the complaint that everyone was not accountable for quality. Now, everyone is accountable, and they know exactly what they are accountable for. Everyone can also see they have a long way to go before they are in a position to point fingers at someone else.

By making everyone's performance visible and not letting problems get swept under the rug, performance measurement systems help keep an organization honest with itself and everyone in it.

As far as the managers and supervisors are concerned, they are now firmly convinced that good performance measures can help improve performance, give them a better understanding of their priorities, and improve their ability to manage and control operations. When asked if they would get rid of the measurement system if they could, their answer was a resounding "No way."

Possible System Improvements

Although the measurement system operates smoothly, some enhancements could be made to it. One would be to use terminals in the plant to enter the data directly and get rid of the written logs. This would eliminate the data entry effort, but would add little to the effectiveness of the system.

Adding the capability to calculate the cost of the quality problems would make the reports more relevant, because frequency alone does not determine the relative cost of defects. Some are more costly than others, and those discovered near the end of a process cost more than those found at the beginning.

Ranking quality problems on the basis of costs instead of frequency would provide a clearer picture of priorities. Knowing the costs would also be helpful in deciding how much to spend on solving particular problems and would help create a greater sense of urgency about eliminating them. Although the plant has made significant progress, there is still opportunity for improvement.

Conclusions

One example is not definitive evidence, but the following conclusions can be drawn from this case since they are supported by similar results in other companies.

1. Measuring the specific problems in a production process greatly facilitates improving its performance by explaining what is affecting performance and identifying where resources can be best applied.
2. If process quality increases, productivity will also increase. No direct attempts to improve productivity were made in this case, even though productivity was measured.
3. Effective performance measurement systems are developed, not designed on paper and then installed. A measurement system must be used in order to identify its weak points, then refined to eliminate them.
4. Performance measures will be a positive motivating factor and a catalyst for change if they are not used to punish people.
5. Accountability is an essential ingredient in making performance measures effective. In the absence of clear accountability, performance measures are not likely to lead to significant benefits.
6. Implementing performance measurement systems requires an initial investment and on-going operating costs, but these costs are much smaller than the costs of poor quality, productivity, and customer service.

An Important Note

In this case, the primary area of opportunity was in-process product defects. Material waste created by the defects was negligible. The only support function that had any significant interaction with manufacturing from a product quality viewpoint was sales, which is included in the defect list.

In the general case, however, manufacturing can be affected by such functions as sales, order entry, product design, plant engineering, maintenance, materials management, and vendors — all of which can be measured.

APPENDIX C

IMPLEMENTING A FORMAL SELLING PROCESS AND PERFORMANCE MEASURES IN A SALES ORGANIZATION*

ABSTRACT

This paper describes implementing a process management framework and performance measurements into a corporate sales organization. It begins with describing the traditional approach to sales management and the potential impact of improving sales performance on revenue and profits. Then, the company's process-based approach to sales management is described along with the key performance measures most relevant for monitoring sales revenue production across sales, marketing, and customer support departments. This case illustrates that viewing sales as a production

---

* This case study is reprinted with the permission of Trailer Vavricka, Inc., which consults with clients to define, document, and sustainably improve their sales process. TVI is currently licensing its advanced *NAVIGUIDE*® software system design for sales process performance measurement and management to leading Sales Force Automation software companies to incorporate into their commercial system products. In parallel, TVI is teaming with selected consulting/education firms to deliver sales production process implementation services and curriculum designed to integrate with *NAVIGUIDE*® enabled software systems.

The company also provides management coaching, and speaks and writes on the subject of Sales Mastery. For a complete outline of TVI services, please contact: Barry Trailer or Joe Vavricka; Trailer Vavricka, Inc.; 319 Shoemaker Lane, Solana Beach, CA 92075; Phone: 619-755-1994; Fax: 619-259-4546; *E-mail:* BarryT@SalesMastery.com, JoeV@SalesMastery.com; *Corporate website:* www.SalesMastery.com

process and implementing process performance measures will enable a company to significantly increase sales and improve sales predictability by increasing productivity throughout the process.

THE TRADITIONAL SALES MANAGEMENT APPROACH

Role of the Sales Force

The purpose of the majority of corporate sales forces is twofold:

1. Keep sales revenue coming into the company at a rate that meets or exceeds budgeted revenue and growth targets.
2. Create customer expectations and relationships which will produce high satisfaction, desire to buy more in the future, and customers who are willing to act as references to influence prospects, generate referrals, and provide feedback that will help improve products and services.

The Traditional Sales Approach

Sales departments traditionally operate informally, that is, without having a formal selling process followed by its sales people. Each salesperson works in his own way, which is a personally derived, non-documented and mostly non-measured approach pieced together from past experiences, training, and ideas gleaned by chance from hearing about other people's adventures. Consequently, management has no way to see how the company's selling function is actually operating as a process.

The common metaphor used to portray the process of sales development is a pipeline. As shown by Figure C-1, a territory's sales opportunities are dispersed along a pipeline depending on the prospect's state of

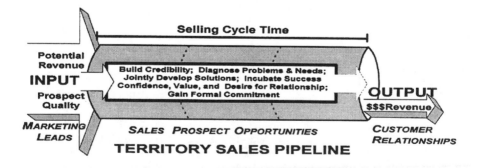

Figure C-1

maturity toward making the buying decision. At worst, the salesperson merely reacts to what prospects asks for and prods them for an order. At best, the salesperson proactively brings the prospect through a joint-effort problem solving experience that builds credibility, confidence, and a desire to commit to a formal customer relationship. The overall cycle time of these types of sales can stretch from a few months to a few years, depending on the product.

At the point the sales opportunity is closed, it passes out of the pipeline and into technical implementation and ongoing customer support, a small portion of which, may be provided by the sales organization. From a sales perspective it is assumed implementation and support functions will follow through to fulfill the customer's expectations and maintain high satisfaction. Customers remain willing to buy more, act as a reference, and provide referral leads only as long as they receive support that meets their expectations.

"Marketing" is the pre-pipeline work to find and develop prospects before they are ready to be entered into the pipeline as live opportunities. The marketing department's purpose is to supply prospects of adequate quality and in sufficient quantity so each territory can meet its monthly sales targets.

Operationally Passing the Buck to Salespeople

Company management normally holds the salespeople ultimately accountable to produce the assigned revenue quotas. However, this amounts to a classic Catch-22 situation. The company's sales revenue production performance results are dependent on multiple departments. But salespeople have no visibility into, nor authority to control or make changes in the performance of the marketing and customer support departments.

Each salesperson's ability to close new accounts or expand business volume from existing ones, is highly dependent upon having highly satisfied customer references available when needed. High quality referral leads throughout the year also help to keep a territory's pipeline healthy and flowing.

The other requirement, of course, is a continuing flow of new prospects into the pipeline from direct mail, telemarketing, advertising, trade shows, and other marketing efforts. If the marketing department's campaigns do not consistently cultivate and unearth enough good quality prospects, then the burden of doing effective marketing work falls onto each salesperson by default. When this happens, not only does this drain time away from selling, marketing suffers because salespeople usually do not have the training, tools, or expertise to do effective marketing.

Nevertheless, traditional sales management often has its highly paid salespeople trying to create marketing wheels from scratch and keep them rolling on their own. Therefore, territories are ineffectively and inconsistently cultivated, leading to an unreliable flow of prospects into the pipeline.

Furthermore, under the tremendous pressure to produce revenues, salespeople resort to working whatever prospects they happen to have at hand — regardless of their low quality. Consequently, the close rate remains chronically low. Desperate to close anything to achieve revenue targets, the company aggressively discounts price, agrees to special terms, makes commitments it can't meet, and gives away normally paid consulting, training, or support services. Ironically, these very situations have a tendency to turn into post-sale nightmares because of high expectations and poor quality throughout the sales production, delivery, and support processes.

Today's sales quotas and market competition are generally much heavier compared to a decade or two ago. Just as a home mortgage for $300,000 @ 9% is a wholly different burden to carry than $50,000 @ 4%, salespeople don't have the capacity to accomplish marketing and customer hand-holding along with their selling responsibility. When they are operating with an unstructured sales approach, it is no wonder that many sales organizations are not as effective as they could be.

Traditional Sales Forecasting Inaccuracy

Sales forecasts are usually produced by each salesperson estimating for each prospect: the potential revenue for each product involved; the probability of closing; and the date it will close. Since every salesperson is estimating from significantly unique perspectives, biases, experiences, understandings, and selling approaches, the company's resulting forecast accuracy is all over the map. Without common understanding of the relationships between inputs, events, and outputs, process outcomes will vary in ways that nobody can predict.*

To make matters worse, upper management then massages these numbers to satisfy any number of political ends and beliefs that have no factual basis. This procedure produces forecasts that are generally regarded by department heads as being so unreliable as to be worthless. This poor predictability significantly handicaps a company trying to plan and structure its operations in order to profit from the business that does materialize.

---

* William Lareau, *American Samurai*, Warner Books, 1992, p. 66.

Hockey Stick Quarterly Revenue Performance

Along with chronically poor sales predictability, inconsistent monthly revenue production is another side effect of traditional "hands off the processes" operating style. In its extreme, pipeline output flow can develop into a recurring "hockey stick" revenue curve. For the first ten weeks of each fiscal quarter sales merely trickle in. Then, 60 to 80% of the quarter's total purchase orders floods in during the final 2 weeks. However, the revenue surge results from a desperation frenzy of unnatural acts — giving away excessive discounts, training, consulting, support, financial terms, etc., in order to close sales.

Executives may be relieved that the revenue number was made, but the workers are repeatedly buried under avalanches of work. Wholly understaffed to properly handle such workloads, quality drops and costs increase. Customers get their first taste of the real relationship — experiencing much less satisfaction than they expected.

Turnover Multiplies Performance Problems and Cost-of-Sales

Under repeated siege conditions created by end-of-the-period sales panics, morale collapses, employees burn out, and increased turnover is inevitable. Sales personnel turnover has steadily risen over the last dozen years. Turnover rates exceeding 30% are common and we have seen many companies with much higher rates. A vacated territory means a vacated pipeline. Its flow will drop, making it necessary for the new person to work hard just to get back to the starting point. This will usually take at least several months. An annual turnover rate of 30% could reduce sales as much as 30% depending on how quickly positions could be filled by qualified people and length of the sales cycle. Add to this the costs of recruitment and training, and the impact on profits becomes very substantial. Unfortunately, the lost profits don't show up in accounting reports and the direct costs caused by turnover are so well accepted and hidden, they are all but invisible.

The Potential for Improving Profits by Increasing Sales Performance

There is enormous potential to increase a company's profit by improving sales performance. As Figure C-2 shows, the increase in a company's profits can be four times the increase in sales, but reducing sales expenses will have minimal effect on profits.

Company P & L Performance Impact	P & L	Sales Expense −5%	Sales Volume +5%	Price +5%
Sales	100.0	100.0	105.0	105.0
Cost of Goods	60.0	60.0	63.0	60.0
Gross Profit	40.0	40.0	42.0	45.0
Mfg. Fixed	13.0	13.0	13.0	13.0
General & Admin.	11.0	11.0	11.0	11.0
Sales Expenses	6.0	5.7	6.0	6.0
Profit Before Tax	10.0	10.3	12.0	15.0
Profit Increase		3.0%	20.0%	50.0%

Figure C-2 The Effect of Sales Expense, Sales Volume, and Price on Profits. (From Hindman, Stephen P. and Sviokla, John J., *Managing Top-Line Computer Applications,* Product #9-192-098, Boston: Harvard Business School, rev. 7-9-92. Copyright 1992 by the President and Fellows of Harvard College. Reprinted with permission.)

Typically most corporations have their sales pipelines operating at only a 10 to 20% close rate. Although this level of performance is not anything to brag about, this low performance level means a relatively small improvement in effectiveness can yield very large increases in revenue.

For instance, the company that can raise its close rate from 20 to 30%, while maintaining all other pipeline performance measures constant, could increase its sales revenue by 50%. Per the above table, this increase in sales volume could create a 200% increase in gross profit performance for the company — with plenty of room to improve beyond the 30% close rate in the following years!

The "Price" column in Figure C-2 shows that a 10× profit leverage can come from effectively increasing prices by giving away less discount, training, consulting, support, and financial terms in order to close sales. Working on higher quality prospects with greater selling effectiveness, stronger references, and less desperation to close, generally softens pressure to discount price, directly raising profit. If the average discount given drops only 2.5 percentage points, it could increase profits 25%!

Improving sales performance can also reduce the sales cycle time. A reduction of average cycle time from six months to five adds two months of selling time which can increase sales revenue over a fiscal year by 20%, assuming all other factors remain the same.

THE COMPANY: The Situation and the New Sales Operating Vision

The Product

The company sells a Total Customer Management®* software system along with implementation consulting, ongoing technical support, and continual enhancements in system functionality. This type of information system is termed an *enterprise wide* application in that users from sales, marketing, and customer support departments all access and update the same customer information system database — integrating all sales, marketing, support, and communication history notes. Having access to the central data, each department can be immediately aware of everything happening with each customer or prospect. The information system helps improve the quality of every interaction with each customer, contributing to the overall quality, depth, and longevity of the business relationship.

Issues in Selling the Product

The company sells the software system through its own sales force, directly to other corporations worldwide. For an information system of this scope, the prospect's buying process is complex. This makes selling a system very challenging. The salesperson must effectively communicate with multiple departments and levels of management including the CEO, CFO, COO, and executives of Sales, Marketing, Customer Support, and Information Systems (IS). Senior users from these departments, as well as consultants who may help evaluate the solution, may also be involved in the buying decision. The salesperson has to identify and convince each involved person of the system's value, and of his company's ability to successfully implement and support the system. References from current customers are strong evidence of that capability, which is why they can make or break a sale.

This complex sale is not merely an exercise in communicating the logical solution to a company's needs. The salesperson must also understand and handle resistance to change. Some of the prospect's employees fear they'll lose heavily if the system were to fail, whereas others think they will lose if it succeeds.

The company faced these same issues in deciding to implement selling as a process into its own operations. Nevertheless, the company became

---

* Total Customer Management® is the registered trademark of ONYX Software Corporation, Bellevue, WA. Website: www.onyx.com

its own best example of operating with a defined process. Using its own product well establishes strong understanding of the system as a business solution, and genuine credibility for the company as a whole.

The Organization

The company had been engaged in a very difficult, two-year, high-growth phase calling for 300%+ growth in the second year, as shown by Figure C-3. At the end of the third quarter (Q3) of Year 1, management decided it needed to implement selling as a process and performance measurement to improve sales performance and control.

Sales Dept.: People & Goals	Year 1	Year 2
Sales managers	3	3
Field salespeople	6	17
Selling experience	5–8 years	5–9 years
Total people selling	9	17
FY revenue goal	$10M	$30M
Sales revenue load per person	$1.1M	$1.75M
Inside sales assistants	4	12
Field system engineers	1	4

Figure C-3   Sales Department Growth Objectives

In Year 1, the direct field sales force had a VP of sales, two sales managers, and six salespeople, all of whom carried sales quotas and direct selling responsibility. All field salespeople except one were remote from headquarters (HQ), working from their homes or small regional offices in their own ways. Four inside sales assistants were put in place to do first pass follow-up contact on all potential prospect leads. After prospects were deemed truly interested and qualified to purchase the product, inside sales transferred them to the appropriate field salesperson, who would take over the remaining steps of the sales cycle.

Operating Issues

In Year 1, the managers and original salespeople were very technically competent with the product's functional capabilities and had intimate product application knowledge. Combined with exceptional selling insight, the core sales group had somehow been able to pull off successive "miracle" finishes each quarter to make the prior revenue targets. This produced a severe hockey-stick revenue performance pattern, which heavily taxed the staff's ability to keep customers satisfied.

Faced with Year 2's 300% revenue growth target and doubling of the sales force, management realized they would have to significantly improve the effectiveness and consistency of how they were operating in order to have a prayer of achieving the new goals. Field sales turnover and the 6-month period needed to get a new field salesperson productive, also had to be reduced. To help shorten the learning period, three field system engineers were added during the first half of Year 2 to team with the salespeople. This was intended to add more technical depth to their selling efforts and make the challenges of the job less formidable to new people.

New Operating Vision

The company decided to bring process structure and performance measurement into the traditionally informally operating sales and marketing departments. The primary objectives were to significantly improve:

- Sales revenue production per person
- Sales revenue predictability (forecast accuracy)
- Management's ability to grow the company's infrastructure while keeping all departments operating as a single team, maintaining 100% of its customers as good references, and preserving the attractive quality of the internal work culture

The executive team believed that to accomplish these objectives, just adding more sales people and marketing resources would not be sufficient. They determined that the company's operational *capability* in producing sales revenue needed to increase significantly and continuously improve to support its growth.

Management believed that given a competitive product/service, and a good corporate reputation, the generation of a company's sales revenue involves a continuous work-flow across Marketing, Sales, and Customer Support departments as shown by Figure C-4. Together they constitute the components of the revenue production system. Optimizing revenue production as a system would eventually call for close coupling and synchronizing of these departments.

Management laid out a summary of their *sales revenue production process*. They captured what needed to be done to continuously initiate,

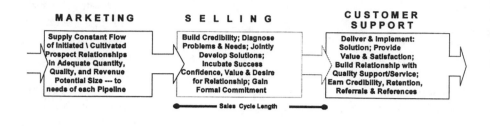

Figure C-4   Department Responsibilities

cultivate, build, and support the company's customer relationships, which in turn produces the company's sales revenue. The diagram (Figure C-5) clearly revealed that this business process was cross-departmental. To name this process they adopted the acronym "CARE" (Customer Acquisition, Retention, and Expansion).

The heart of the CARE revenue production process is the sales pipeline. It contains an ever-changing volume of potential sales revenue, in the form of individual sales opportunities as they flow through the selling

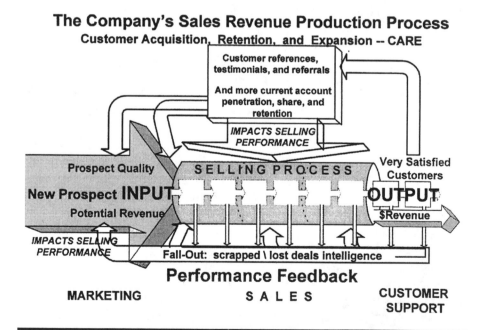

Figure C-5   The Care Process. (© Copyright 1993 Trailer Vavricka, Inc. Solana Beach, CA 619-755-1994. All rights reserved.)

process. The never-ending challenge facing every corporation is keeping its sales pipelines flowing at full capacity with good quality prospects. Each week prospective sales opportunities can and do scrap out of the pipeline anywhere along their way as "fall-out", for a myriad reasons. New and different sales opportunities are being entered as "input." Others are closed as booked business, moving out of the pipeline as "output." The volume and content of "sales-in-process" is in constant flux. Some prospects are maturing, moving forward in the selling/buying process, while others stay put or even move backwards.

From the CARE model, the responsibilities of the field sales staff and their corresponding performance measures were determined. Each field person would be accountable for getting their pipeline's operating performance measurements healthy and keeping them in balance. Sales managers, in addition to improving the performance of their consolidated regional pipelines, would be accountable for improving and synchronizing cross-departmental activities to attain it.

All managers, including the VP of Sales, were to have the same operating performance measures as salespeople in their territories. This made the performance picture consistent over the whole sales force. The ability to compare performance between territories and to the sales force average, provides an ongoing objective basis for managers to identify where they need to focus their coaching efforts. Everyone will see the graphs and differences in marketing and selling process behaviors. By identifying what methods work better than others, management can systematically improve revenue production and reduce overall variation in performance. This will result in both increased revenue and better revenue predictability.

This unified picture of operating performance would serve as the driver for organizing and synchronizing marketing's functions as supplier to the territory pipelines. Marketing will be able to determine future prospect requirements and will be called upon to systematically improve its lead generating effectiveness. The system will provide the sales history database from which marketing can get feedback on wins, losses, lead sources, industry segments, competition, and other prospect parameters.

To manage the all-important customer references, customer support would regularly update a "Customer Reference Availability" database, while sales and marketing would likewise update the database on all uses of customers as a reference.

To begin to significantly increase revenue, management first needed to determine how well the operation was currently performing. After that, it would need to see the measurements on a regular basis in order to keep the revenue production system components synchronized.

IMPLEMENTATION: The Selling Process and Performance Measures

The Selling Process Map

A "customer relationship" is the result of two or more independent processes combining and groups of people interacting. Two processes always present in any customer relationship are the selling process of the vendor and the customer's buying process. Other processes can be involved if other parties are involved in the relationship for buying, selling, implementing, or supporting the products.

Unlike manufacturing environments that are set up to control the entire fabrication process, Sales can only try to influence the actions of the prospect. Even in executing the fundamentals of its selling process, Sales must accommodate the needs of the prospect's buying process.

This points to why many veterans insist that selling is an "art" and that "talent" is what sells. Insofar as art is a medium to convey knowledge, meaning, and emotion, salespeople indeed educate and create interpersonal rapport, trust, and confidence with prospects. They must certainly inspire emotional desire in the prospect to *want to* commit to a relationship.

Yet, as a repeating business operating function, there is also mechanical process or science underlying selling. A sales manager once described selling as being like a bicycle. The gears, chain, sprockets, pedals, frame, and wheels are the process — the bicycle's capability and capacity. However, the steering, balancing, and quality of energy powering it — that's talent. The selling process map is a common framework for guiding and measuring selling workflow through the sales pipeline. Development of the map started with Trailer Vavricka, Inc. facilitating a two-day session for the sales force to map its own "best-methods" selling process. This identified and captured what seemed to produce the most consistent closing and best quality of customer results.

The sales force defined six macro-steps to their selling process. Each step contains an objective (*what* it is designed to accomplish), a desired result (*how* to determine the objective is accomplished), and two subsets of detail step actions that usually lead to achieving the desired result. One set of actions list what the selling team generally needs to accomplish, while the other set lays out what the prospect's buying team typically needs to do at each point in the cycle.

In the same session, the group also derived their prospect quality criteria, set its team operating rules for using the process consistently, and loaded their pipeline database with all current sales-in-process. This produced a unanimously agreed upon selling process in sufficient detail to serve as the common structure to meaningfully measure operating performance for each individual territory and every management level.

Sales Operating Performance Measures

Performance measures can show how effectively the combination of process and talent is operating as in Figure C-6, which shows pipeline revenue input rate. The performance measures can also identify significant changes in performance as they start to occur and the pipeline's bottlenecks. With such early warning signals, sales, marketing, and support managers will be able to take timely action to keep the revenue production in control and more predictable.

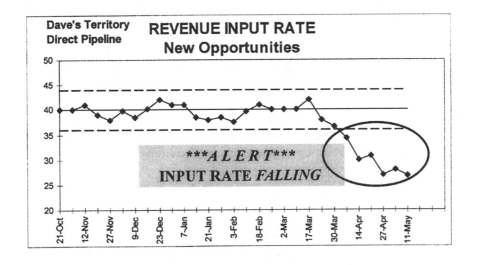

Figure C-6   Pipeline Input Revenue Rate — Percent Full

Managers reasoned that implementing sales process performance measures would enhance four fundamental management capabilities. They would provide all three CARE departments with quantified operational feedback, fact-based understanding, and clear requirements for:

1. Maintaining the production of sales revenue on-target to current FY budget goals;
2. Increasing sales production capacity *at least* a sales cycle length ahead of when increased sales targets call for greater monthly output;

3. Continuously improving the processes within each CARE department, as well as the daily execution of them, to increase operating *capability*;
4. Synchronizing the CARE departments to operate as a single team, with the common goal of optimizing the overall operating performance of the sales revenue generation system.

Sales process performance measures would also directly support the salesperson, who's role was defined as a Territory *Business* Manager, responsible for:

1. Producing 100% or more of the current FY's assigned quarterly revenue targets;
2. Developing a continuous business pipeline flow adequate to meet the monthly revenue targets across fiscal quarters and fiscal year boundaries — with 100% customer satisfaction;
3. Improving quarterly forecast accuracy to within ± 15% of actual.

Due to the dynamically changing status of each pipeline's performance and capacity, everyone's measures need to be recalculated each week, in order for the feedback graphs to effectively reveal changes in performance. The following measures were selected to monitor the performance of the sales process with respect to the given objective.

Producing 100% or More of the Quarterly Revenue Targets

- **Input revenue flow per month** is the amount of new revenue needed by each pipeline, and would be calculated according to how the territory's pipeline was actually operating. For example, a pipeline operating at a 20% close rate with a $100K/month quota would need $500K of new input each month to keep its flow going; but only $250K if its close rate had improved to 40%.
- **New sales project quality** is another key leading performance indicator, along with input revenue flow. Better quality prospects typically close at significantly higher rates, taking less average cycle time. The opposite is true for poorer quality prospects. To develop a consistent way to measure the relative *quality* of each prospect opportunity, salespeople agreed on the top five most important determinants of a prospect's quality. These criteria went beyond merely arbitrary demographics into the characteristics most germane to forming a long-term customer relationship. A rating scale of –5 to

+5 was used for each criteria. This measure showed the average quality of the input stream as well as the overall pipeline contents.

- **Close rate by process step** is the calculated probability to close from each step of the sales process, for each pipeline. This is also used in calculating several other measures such as input rate, pipeline percent full, and projected output.

- **Pipeline percent full** shows whether the revenue of sales-in-process is enough to support the monthly revenue objective. This is a good indicator of how healthy the pipeline is according to its current close probability, cycle time, and its volume-in-process capacity. Figure C-7 is an example of a performance chart.

- **Average project revenue size** is calculated across all the sales projects currently in a pipeline. Revenue growth can be significantly increased by the pipeline's population of prospects becoming larger in average size, requiring a lower total number of projects to be found and worked. In the company's case, it takes almost the same amount of effort and time to win a small contract as it does for a large one.

- **Output revenue percent to fiscal quarter target is** simply the actual revenue recognized per fiscal quarter as a percentage of the territory's revenue quota for the same quarter. This is the traditionally used results measure.

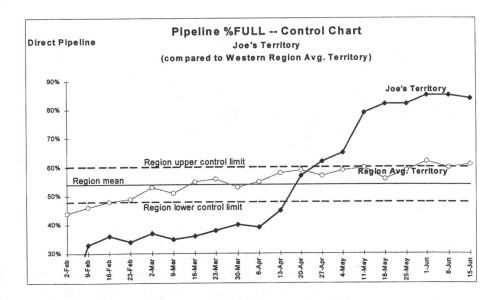

Figure C-7   Sample Performance Chart

Developing a Continuous Adequate Pipeline Flow

- **Pipeline forecast projection by month and YTD cumulative position** shows the salesperson and manager how the current pipeline contents will probably flow out in future months and accumulate in YTD revenue. This is based on the individual performance measures of each pipeline, giving each person a relevant picture of where they stand and are likely to be in future months. Everyone can see whether their projected sales revenue position is below or above their target. The projection spans fiscal period boundaries to keep visible the need to maintain continuity in each pipeline's adequate flow at all times.

- **Cycle time by process step** is the total time in weeks that the average sales project takes to go through the pipeline's sales process steps. Steps with large time sinks should be candidates for investigating what is causing delays to see if the process can be improved. Cycle time is also used in calculating several other measures such as projected revenue output, pipeline %full, and course correction. A sample chart is shown by Figure C-8.

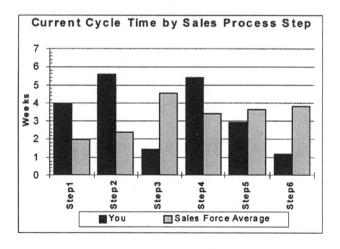

Figure C-8   Selling Cycle Time Comparison Chart

- **Percent fall-out by process step** shows the portion of potential revenue of sales projects that were lost or otherwise scrapped out of the pipeline. Large portions of fall-out early in the sales cycle usually points to low or misgauged prospect quality, whereas late in the process, it suggests a need for selling process or execution improvement. Reducing fall-out has the effect of raising the close rate and reducing overall average cycle time. Moving fall-out forward in the sales cycle prevents wasting time and resources on future scrap, enabling them to be applied to better opportunities. This can raise the close rate and improve sales volume. A sample chart is shown by Figure C-9.

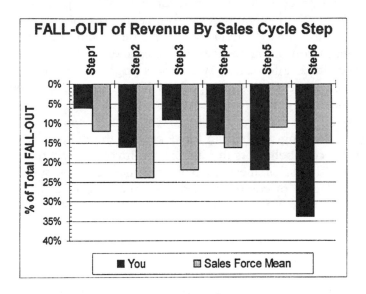

Figure C-9   Revenue Fall-Out By Process Step

The fall-out in each step can also be analyzed by competitor, product, industry segment, and selling team. The patterns of fall-out can provide insight into what is causing it and how it might be reduced. Similar analysis can be done on closed opportunities to see where particular people or approaches are stronger, so those tactics can be deployed to all territories.

Improving Quarterly Forecast Accuracy

- **Quarterly forecast accuracy** is measured by the difference between forecast and actual revenue as a percentage of the forecast. The forecast is what was predicted as of the last day of the previous fiscal quarter. Each quarterly period is calculated each month on a rolling quarterly basis.

Implementation: Consistent Daily Use and Feedback

Management consistently encouraged all the salespeople to adopt and regularly execute the sales process procedures as a daily habit. The goal was to establish the process as a standard framework for managing sales opportunities as individual projects. Management felt this framework would accelerate getting every salesperson up to a high level of selling competency and provide managers with the information they needed to become effective coaches. Managers reinforced the process orientation by using the sales process as the context for every sales opportunity discussion, resource and quotation request, forecasting, planning and debriefing every sales call, and all interfacing of the field with marketing, inside sales, technical consulting services, and customer support groups.

The sales force information system was upgraded to incorporate the mapped sales process steps, prospect quality criteria rating, pipeline per territory, and to regularly capture the needed data on all sales project movement through closing or fall-out. The system would serve as the company's sustaining mechanism for deploying the cultural continuation of its best process behaviors throughout the involved departments. Management felt the sales measurement and management system would provide the degree of control it needed — and had never had before.

The sales management system needed to have algorithms built into it to make the performance measurements each weekend for each remote salesperson, up through the consolidated VP Sales level. The system would have to:

- Measure the chosen vital signs of *in-process* sales revenue production flow, and show statistically significant changes and forming trends in actual operating performance
- Provide on demand the sales forecast projection based on the actual operating statistics as measured each week
- Report the measures graphically so everyone could easily understand and monitor performance

Due to limited internal programming resources, implementing a complete sales management system with these capabilities was going to take much longer than management desired. To get started as quickly as possible, an interim system was put in place to measure the pipelines for current territories and the aggregate national pipeline. Tools such as Excel's database query, pivot table, and graphing were used to summarize the data and track changes in the operating measures. Although not as sophisticated as the new sales management system would be, the patchwork system was capable of providing reliable performance measures.

Each remote salesperson had a computer, which would automatically send all sales project updates made each week to the central database. Salespeople made up a common set of system updating procedures so they all understood how to consistently use the sales process and do timely and accurate system updating as sales projects progressed. This made the data very complete and reliable.

Impact On Operating Performance

The sales operation was able to increase both its headcount and output rate over the first two quarters of Year 2. The YTD performance measurements of Figure C-10 show an interesting picture of current sales operating conditions and reveal the challenge for attaining the next two quarter's sales targets.

Operating Performance	Year 1 (end 4th Qtr.)	Year 2 (end 2nd Qtr.)
Input Prospect Quality (–5 to +5)	–2	0
Pipeline %Full Next Cycle Length	33%	60%
Sales Cycle Length	6.1 mo's	5.0 mo's
Close Rate Field Sales Pipeline	14%	31%
Average Project Revenue Size	$73,000	$98,700
Output Revenue % of Goal (at a 300% growth rate in Year 2)	101%	87% YTD
Relative sales level	100%	240%
Quarterly Forecast Accuracy	+/–58%	+/–11%
Turnover in Sales Personnel	33%	9%
New Salesperson Ramp-up Time	6 mo's	4 mo's
% Customers Referenceable	99%	99%

Figure C-10   Impact On Performance — From Year 1 To Year 2

The most striking improvement is the close rate jumping from 14% in Year 1 to 31% for the first two quarters of Year 2. This was partially due to improved selling effectiveness, resulting from everyone using the "best methods" selling process. Also, the executive management team relentlessly assisted the salespeople in actively meeting with prospects by traveling to prospect sites and hosting prospects during their visits to the company. However, Q3's revenue target is 40% higher than Q2's. Besides that, counting on maintaining the 31% closing rate without improving selling capability in some fashion was risky. Accordingly, sales management began formal sales opportunity reviews and initiated planning for all above average size sales projects that moved into the solution demonstration phase of the selling process.

Although new field sales people are quickly becoming competent to manage system sales as a result of the structured selling process and better training, they are having difficulty keeping their pipelines full. The company's initial effort to fill the pipelines was centered on beefing up the inside sales group's capacity. They were being counted on to find and follow-up more of marketing's general leads in order to stock the new territory pipelines as new salespeople were hired. While this helped fill the national pipeline, all the newer territories were far behind their objectives going into Q3.

Average prospect quality, cycle time, and project revenue size have also improved. Another important contribution to performance improvement is through the reduction in turnover. When turnover is low, the territory development effort can be continuous with most management time being dedicated to coaching instead of repeated hiring and training.

Quarterly forecast accuracy has greatly improved so far for Year 2. However, the hockey-stick nature of more than 70% of the closed revenue coming in the last month of the quarter, is restraining management's confidence in overall revenue predictability. It is also costing the company in having to discount, provide special terms, and give away normally free training and services to get orders closed before the end of the quarter. There has been a large amount of slippage in expected close dates and revenue (additional operating performance measures), but this appears to be more of a problem with estimating than with operations performance or customer behavior. Rules for the sales force to consistently estimate project revenue as well as close date, and close probability are being instituted to decrease the estimating errors.

What is now crystal clear to management is that new territory marketing cultivation has a 1-2 sales cycle lag time (5 to 10 months) before it can establish an adequate input flow of qualified prospects. In Q1 of Year 2, Marketing started a rudimentary on-going contact pattern for any prospects field salespeople or inside sales provided to them. They are now starting Q3 with resources to launch support for larger scale, multiple-contact-points-per-prospect cultivation of each territory. Unfortunately this will take about the next two quarters to really kick in. Until then, the salespeople will have to employ guerrilla prospecting tactics in an all out effort to get enough prospects into their developing pipelines to make revenue targets for this year and be in position to start next year at full pace.

Some specific instances where the performance measures were instrumental in initiating corrective or other actions to improve performance are given below.

1. At the end of the 4th quarter of Year 1, the Sales VP saw that the revenue projection based on the system's calculated performance data showed that the next two quarters were going to fall far short of objectives. This was in contrast to his own estimates that were based on a much rosier picture. He called an emergency meeting with the Marketing and Inside Sales managers to develop a plan for keeping the pipelines full with a steady stream of new prospects. They worked out an approach to target the top 50 potential customers in each territory for sustained cultivation — combining specifically designed direct mail, seminars, and teleselling programs. This increased the Pipeline Percent Full measure 20 points, making achievement of the sales goals at least a realistic possibility.

2. The Inside Sales group contacts raw sales prospects to determine which are worth passing on to the field. In an effort to increase the number of prospects getting into the pipelines, the Inside Sales manager put incentives in place in some groups. The incentives were based on the number of prospects each Inside salesperson passed on to the field. Two weeks later, the pipeline showed a marked increase in prospects, but prospect quality, as assessed by the salespeople, also showed a sharp decline. Quality was apparently suffering in favor of quantity. This was confirmed by a more detailed analysis, which showed the drop in quality was happening only in the areas that were under the incentive plan.

   The incentive plan was subverting the primary purpose of Inside Sales, which was to screen out poor quality prospects so they would not waste the field salespeople's time. To prevent this, the incentive plan was changed to include prospect quality criteria and subtract prospects rejected by the field as poor quality. Prospect quality quickly got back to normal with little decrease in prospect volume. Had the quality measures not been present, it undoubtedly would have taken several months for the quality problem to become evident.

3. At one point, the Southern Regions' close rate was 14% lower than the average. Investigation showed the largest cause of the problem was the quality of customers that could be used as references and for getting referrals into other companies in the region. Surveys of the key accounts for referrals revealed that most of them were only using about 20% of the product's capability, which was

certainly not resulting in highly satisfied references and great referrals.

A customer CARE team was formed to raise the customers' level of understanding of the system's scope and everything that it could do. As a result, the key customers began additional training and implementation efforts, quickly realizing significantly greater benefits and return on their investment. In return, they became very strong references and referral generators for the Southern Region, speaking to prospects one-on-one and through organized sales seminars. This increased the region's overall close rate by 17% in the ensuing 5 months.

4. A Western Region sales rep had the largest amount of pipeline fall-out, most of which was happening on the last step of the sales process. He also had the lowest closing rate of anyone in the region. The sales rep and the regional manager examined what was happening and concluded that the primary cause of the problem was poor execution in getting to upper management in the selling process.

   Querying the system, the regional manager then identified other salespeople across the company who had high close rates and the least fall-out in the last steps of the pipeline. Along with the other regional managers, they were then surveyed to determine what techniques for getting to and selling to upper management worked best for them. This included telephone scripts, questions asked of customers, benefit statements, article reprints, meeting agendas and formats to run meetings.

   In 6 months, the sales rep's close rate was above the regional average and his last step fall-out was reduced 70%. In this case, the performance measurement system not only identified a specific problem; it told management where to look for the solution. Without the performance information being available, identifying and correcting the problem would have taken much longer — probably at least another year, if ever. The sales training manager is organizing this into an education module for the regional managers to present to all salespeople to improve everyone's performance.

Impact on the Organization

The company's CARE departments, along with the executive committee, have become galvanized in their effort to contribute to the company achieving the needed operating performance improvements. Everyone is focused on finding ways to improve their individual performance in order to support the team effort.

The performance measurements have been instrumental in helping management to understand and zero in on what areas in the CARE process would contribute most to improving the company's revenue production performance. Having the common, fact-based picture of what is happening in sales operations has enabled more rapid concurrence on deciding what needs to be done, priorities, and who needs to do it.

Both inter-department and management-worker unity has noticeably improved as a result of having common objectives, clear responsibilities, and reliable, objective performance information. Departments are operating as a team and all members are playing with heart. The quality of the culture is itself improving — even amidst the demands, strains, and pains of rapid growth.

As investments in continued process improvement, management decided, among other items, to:

- Build up the Inside Sales infrastructure and hire a manager to analyze the process and continuously improve its capability to provide qualified prospects to all field pipelines.
- Utilize Inside Sales as a filter to prevent field salespeople from wasting their limited selling time and resources running after sub-quality prospects — which salespeople tend to do when that's all they have available.
- Hire vertical market industry experts to build specialized marketing plans, produce education programs for field personnel, develop customized solution demos, and generate other initiatives.
- Implement sales opportunity reviews and planning sessions where a salesperson and manager will formally review sales projects that are near closing to develop comprehensive selling plans to win the business.
- Develop a customer reference system to enable quick identification of available references that match the prospect's profile and track reference usage, actual involvement, and effectiveness.
- Add more field system engineers to team with salespeople in custom demo modeling of prospects' selling, marketing, and support processes.

A considerable operating benefit of the measurement system will be better predictability of when and what business will be materializing. Armed with this information, department managers will be able to make better staffing and training decisions further ahead of when the business will actually materialize. This is how the company sees it will be able to better maintain the quality of the internal work environment and 100% customer satisfaction.

Potential System Improvements

The information system currently has several additional enhancements underway to add other key performance measurements and expand the reporting capabilities to include control charting of all the weekly performance measurements. New measures will include:

- Sales process step and step action execution tracking across sales projects
- Close date and revenue slippage
- Rolling 1 to 6 month forecast accuracy
- Rate of flow within each pipeline
- Project age tracking to identify where action/scrap/recycle decisions must be made
- Correcting additional input revenue amounts needed each month to compensate for projected revenue shortfalls in each territory

Control charting the weekly performance measurements helps put each week into the context of all the previous weeks. Seeing the series of performance points graphed over time shows the natural band of variation around its "normal" average. In the same way for each territory and manager's pipeline, control charts detect and give exception ALERT signals for taking timely corrective action.

There will also be better ways to analyze closed and fall-out projects for finding the largest areas for potential improvement and identifying where better methods are emerging which can be used to improve performance.

Marketing will begin to directly help improve field close rates for each product line by looking across the territories for standout patterns in effectiveness against particular competitors in specific industry segments. Marketing will then collect the best methods, documenting where each fits best within the standard sales process. These changes will then be sent to all salespeople as they begin pursuing opportunities fitting those profiles. Product line performance data will also help better manage the decision process for introducing new products and deleting non-productive ones.

Conclusions

Although there are many improvements to be made to the sales performance measurement and management systems, some firm conclusions can be drawn from this study:

- Selling is a production process that can be measured. It may not be possible to measure sales as precisely as a manufacturing process, but the process can be measured well enough to give management relevant, useful, and timely information for making operating and strategic decisions.
- The performance measures that were developed made a significant contribution to improving performance throughout the sales process. Management had to make the decisions and take action, but there is no question that the performance measures were effective in identifying problems on a timely basis and also in helping to solve them.
- Using a structured process framework and performance measures had no negative effects on performance or morale. Instead, the effects were all positive. Most of this can be attributed to the leadership of the company's top management, but this experience illustrates there is nothing inherently objectionable about measuring individual and group performance in sales and marketing.

APPENDIX D

GLOBAL WARMING?*

Note: This article is an excellent illustration of how statistical analysis of data can be misused to draw incorrect conclusions. It also illustrates that it is usually not difficult to find a short-term sample of data that will support any argument that anyone cares to make. Please note that the conclusion drawn by the author is not that global warming doesn't exist, only that the described data provides no evidence of a trend.

Statistical Analysis

Global warming is a theory in search of supporting data. However, in the search for supporting data, we should avoid misinterpreting our data.

In the July 4, 1996, issue of *Nature*, Santer et al. use data such as those in Figure D-1 as evidence of global warming. The values shown represent the annual average air temperatures between 5,000 feet and 30,000 feet at the midlatitude of the Southern Hemisphere. The zero line on the graph represents the normal temperature, and the values plotted are the deviation from the norm for each year.

The data of Figure D-1 show a clear upward trend between 1963 and 1986. However, when we fit a regression line to data, we are imposing our view upon the data.

If we know of some cause-and-effect mechanism that relates one variable to another, then regression lines are appropriate. But does the *year* cause the trend shown in Figure D-1? While regression equations are useful in showing relationships, these relationships may be either causal or casual. At most, the relationship in the figure is casual.

---

* Copyright © 1997, Donald J. Wheeler, Ph.D., Statistical Process Controls, Inc., Knoxville, TN 37919. Used by permission. All rights reserved.

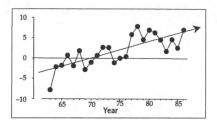

Figure D-1

But is the trend in Figure D-1 real? Or is it just noise? This question can be answered in two ways: Check for internal evidence of a trend with a control chart, and check for external evidence of a trend by adding more data as they become available. Figure D-2 shows these 24 data on an X-chart. The values for 1963, 1978, 1980 and 1986 all appear to differ from the norm (i.e., they are outside the limits). Hence the "cool" year of 1963 combined with the "warm" years of 1977 through 1986 do suggest a possible trend. So there is some internal evidence for a trend in these data.

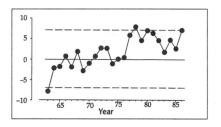

Figure D-2

The limits can be adjusted for this possible trend in the following manner. Compute the average for the first half of the data. Years 1963 through 1986 had an average of –0.74°C. Plot this average vs. the midpoint of this period of time — halfway between 1968 and 1969. Compute the average for the last half of the data. Years 1975 through 1986 had an average of 4.55°C. Plot this value vs. the point halfway between 1980 and 1981. Connect these two points to establish a trend line.

The distance from the central line to the limits in Figure D-2 was found by multiplying the average moving range by the scaling factor of 2.660. The average moving range is 2.663°C. Thus, limits will be placed on either side of the trend line at a distance of: 2.660 × 2.663°C = 7.08°C (see Figure D-3).

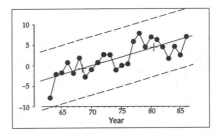

Figure D-3

Thus, the internal evidence is consistent with a trend for these data. But what about the external evidence? Professor Patrick Michaels of the University of Virginia added six prior years and eight following years to the data of Figure D-1 (see Figure D-4).

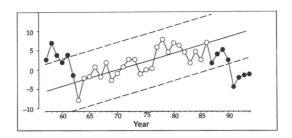

Figure D-4

So, if the data for 1963 through 1986 are evidence of global warming, then the subsequent data show that we solved the problem of global warming in 1991. However, if the interpretation of the data for 1963 through 1986 is merely wishful thinking, then we may still have some work to do.

The 38 values are placed on an X-chart in Figure D-5. Once again, the central line is taken to be zero in order to detect deviations from the norm.

Thus, while 1963 was cooler than the norm, and while 1977 through 1990 were detectably warmer than the norm, there is no evidence in these data to support the extrapolation of the trend line shown in Figure D-1. Obviously, there are cycles in the global climate, and any substantial evidence for global warming will require a much longer baseline.

The first principle for understanding data is: No data have meaning apart from their context. We cannot selectively use portions of the data to make our point and ignore other portions that contradict it.

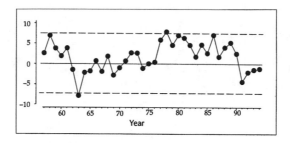

Figure D-5

The second principle is: While all data contain noise, some data may contain signals. Therefore, before you can detect a signal, you must first filter out the noise. While there are some signals in these data, there is no evidence of a sustained trend.

About the Author

Donald J. Wheeler is an internationally known consulting statistician and the author of *Understanding Variation: The Key to Managing Chaos and Understanding Statistical Process Control, Second Edition.*

APPENDIX E

MEASURING AN INVENTORY TRANSACTION REPORTING PROCESS

ABSTRACT

This example of operational performance measurement illustrates how a seemingly unmeasurable process can be measured. It demonstrates that logical relationships between variables exist in any process and when the details of these relationships are understood, it will be possible to develop performance measures for the process. The process in question is the reporting of inventory transactions in a job-shop fabrication and assembly plant.

The Situation

The company made valves and pumps for controlling steam and liquids. Many of the products were quite complicated and expensive. Since many different product options were offered, such as capacity, types of fittings, features, and materials to withstand certain environments for the products, over 30,000 different items were kept in stock.

This alone made inventory control important, but with many of the parts being costly and made or purchased in small quantities, there was little margin for error. When inventory records were not correct, a shortage of even a relatively inexpensive part, could cause some or all of the following problems:

- Interrupted production and late shipments
- Increased work-in-process inventory

- Expedited shipment or production of parts at significantly increased costs
- Substitution of more expensive parts
- Added labor costs to work around the problem
- Increased likelihood of quality problems in the final unit
- Cannibalization of other in-process units, creating a whole new cycle of all of the above problems

For these reasons, inventory accuracy was critical, but the inventory system was not performing well in that regard. More than half of the larger and more complicated units in production would experience one or more shortages that would cause serious problems. On less complicated items, about 20% of the production orders would experience an inventory shortage, which is poor performance by any measure. In addition, much time was wasted by the inventory control staff physically verifying parts were available because there was no confidence in the inventory data.

Most of the inventory transactions were captured automatically and were error-free, but it was possible for some transactions to be omitted because of human error. About 15% of transactions, however, required manual input. These transactions were created by incidents such as changes to orders, design changes, bill-of-material errors, defective parts, and stock-pulling errors. An analysis of inventory discrepancies indicated the manual transactions were a major source of errors, but there was no way to determine the extent of the problem. In essence, this was the problem: identifying and measuring transaction errors occurring in the process. If this could be done on a timely basis, it would be possible to determine what was causing the errors and take action to eliminate them.

Determining What to Measure

Determining what to measure started with preparing a detailed flowchart of the inventory control system, identifying all inputs, outputs, and sources of transactions. All transaction sources were then investigated to determine the different types of transactions that were made at each point and the reason for the transaction. Variables such as the quantity, part number, inventory control parameters, and work group making the transaction were also captured. Detailed records were kept of all manual transactions for two weeks. Then the data was analyzed to determine if there was a logical structure to the transactions or relationships between the variables.

As expected, there were some relationships between the process variables. For example, the quantity of parts withdrawn to replace defective units would normally be less than two, but the quantity needed to accommodate a change to a product specification would be in the range of 10 to 20. There could always be exceptions, but the quantity of parts in any transaction would be in a range determined by the following variables:

- Transaction type — withdrawal or return to stock
- Reason — replace, bill-of-material error, design change, incorrect quantity issued, etc.
- The operation making the transaction
- The quantity of parts normally ordered (purchased or manufactured)
- Inventory class of the part: normal stock, make-to-order, only ordered if master level unit is ordered, bulk, or product development

Other logical relationships were also identified within the inventory system, which could provide useful feedback and measurement data. For example, if a withdrawal resulted in the "quantity on hand" going negative, something was obviously wrong. Similar logic applies to the "quantity committed" and "quantity on order." Balance checks could also be made between variables in the system such as outstanding purchase and production orders and what is reported as delivered to stock. All transactions and variables throughout production control, inventory control, and purchasing could be similarly linked so that any imbalance or illogical transaction could be identified.

Implementation

The relationships between the variables in the inventory, production, and purchasing systems were used to construct logical decision rules to screen every reported transaction. Any items violating the rules were identified as a probable error and literally kicked out of the system. The rejected transactions would then be validated or corrected and reprocessed. An override code was used to bypass the logic checks for the transactions that exceeded normal limits but were valid.

The percentage of illogical transactions provided a measure of the overall reporting performance of the inventory system. The quality of the transaction reporting process was measured as the percentage of total transactions that failed the logical tests. Accountability for the errors was established by summarizing the errors according to the work group making the transaction. Here again, decision rules were used to properly identify the responsible work group.

When the system was first implemented, every transaction reported the previous day and rejected by the system was promptly investigated to determine the likely cause of the problem. With an initial error rate of 5 to 6% and about 3,000 transactions per day, this meant investigating 150 to 200 items every day. Needless to say, this initially took a considerable amount of time, but as more was learned about the system and the errors being made, the time required to do all the follow-up rapidly dropped to a manageable level of one hour or less each day.

One of the first things learned from this detailed follow-up, was that the decision rules needed to be modified. In some cases, the rules were too strict, resulting in valid transactions being reported as errors. In other cases, the gates were too wide, letting too many errors slip through undetected. After a few weeks of refining the rules, they became very effective filters. No detailed analysis of the decision rules' effectiveness was conducted, but they became about 102% effective, letting almost no errors slip through and only reporting a small percentage of valid transactions as errors. The filters were purposely designed to err on the side of rejecting valid transactions rather than letting errors pass through.

Using the Performance Measurement Information

A very significant initial finding was that most of the errors were caused by confusion about how to report certain transactions — everything from what transactions must be reported to how the manual transaction form needed to be filled out. No instances were found where a person was deliberately careless about reporting, although that was the assumption of some managers. It was a classic case of the fault lying in the process, but the blame being placed on the people.

The two major shortcomings of the inventory control system were the complexity of the form being used to record miscellaneous transactions and inadequate physical control of inventory. The original form was complicated, crowded, and confusing. To an outside observer, it looked fine, but to a person using it, the form looked like something designed by a committee of near-sighted Washington bureaucrats. Every possible transaction could be reported using the form — that is, if anyone could figure out what data needed to be entered and where to put it.

This problem became evident from investigating the reporting errors that were identified. Input from users made the problems with the form plainly evident in the first two weeks. By then, it also became apparent that in any work group, only a small set of all the possible types of transactions were ever used. Consequently, the decision was made to simplify the process by replacing the general-purpose form with special forms for each type of transaction and/or work area. The forms were designed so that every data field and character that needed to be entered had its own box or space. Furthermore, there was no room for entering any data that was not needed. For example, raw material forms provided one space for entering tenths of feet that had to be filled in that was preceded by a decimal point the size of a pencil eraser so it would not be overlooked. Anyone using the form could easily see what data was needed and if any data was missing. Although going from one form to fifteen may not seem like simplifying anything, it was a great improvement from the users' perspective.

The problem of the physical control of inventory did not become evident directly from the performance measures or data, but from spot-checks to assess accuracy of the inventory figures. When the physical count of parts was different from the inventory record, the count was always lower than the inventory number, except for a few cases where receipt of goods was not reported. The differences were rarely large, but they were still important. This meant there was some "leakage" in the system.

Observation of the work area soon revealed that small quantities of parts were being liberated from inventory by people from the assembly area when the stockroom staff was on break or otherwise not present. This was being done out of necessity when something was needed to complete a job on the shop floor and any delay was not felt to be acceptable. Although forms were available, the data needed to fill out the form might not be readily available and intentions to "do it later" would be forgotten.

While failure to report some of these transactions was due to human indifference, the system to get the required parts was cumbersome and time-consuming for the people on the shop floor — especially when being pressed to meet a schedule. Accordingly, the system was "reengineered" to provide proper physical control and make it efficient for the plant personnel. The key features of the new system were:

- **Restricted access** — All doors to the stockroom were welded shut except one for deliveries and another for withdrawals, which were locked unless an inventory person was present.
- **Added staff** — A clerical person was added to the stockroom to provide continuous coverage and to assist in processing all the inventory and production paperwork coming from the plant.
- **Revised reporting procedures** — Most of the paperwork burden was removed from the plant personnel. Much was eliminated and most of what was left was transferred to the stockroom clerical assistant.

There was considerable doubt that the new system would work. The plant personnel were sure it was going to slow them down. After a few weeks, however, the new system was working very smoothly. It was faster, easier for the plant personnel, and 100% accurate. In fact, everyone was asked if they wanted to go back to the old system, but there were no takers.

Results

As shown by Figure E-1, transaction errors were reduced from 5% to less than 0.2% in six months from the start of the project and in slightly over three months after the measurement system started working. The small number of errors (0.1 to 0.3%) being reported in the final weeks shown on the graph remained at that level thereafter. Almost all of these apparent errors were valid transactions that exceeded the tight filters in the system. This meant the inventory, purchasing, and production reporting systems were essentially error-free. This was confirmed during a year-end audit, when the auditors found no discrepancies between inventory records and random physical counts of items. In fact, one auditor inferred that the inventory figures were somehow being manipulated, until it was pointed out that would require forcing him to select specific items for spot checks by mental telepathy. The inventory numbers were so accurate, a fortune could be made by making bets and giving fifty-to-one odds that the number of parts found in stock would exactly match the inventory records (except for bulk items, of course).

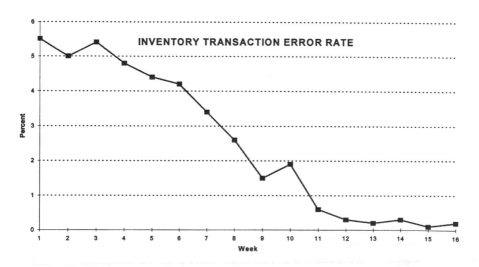

Figure E-1   Improvement in Reporting Inventory Transactions

Productivity in the inventory control department also improved about 20%. The person added to the stockroom was transferred from elsewhere in the department, so there was no net addition of staff. The improved efficiency of the system, primarily through reducing errors, enabled one other person to fill a vacancy in another department. The staff could have been reduced by at least one more person, but management decided that it was better not to lose knowledgeable people and to let growth absorb the capacity released by the increase in productivity.

Accurate inventory records also increased productivity in the plant. More accurate planning and a dramatic decrease in stockouts made a noticeable difference in all of the assembly operations. Unfortunately, no productivity measures were in place, so the improvement can only be estimated to be in the range of 10 to 20%.

Conclusions

The most significant conclusion derived from this case is that it was possible to measure the quality of the transaction reporting process. Although it was initially not clear that this could be accomplished, when the details of how the process worked were understood, it became apparent that the logical relationships between variables provided a way to measure reporting accuracy.

The second conclusion, is that being able to identify where problems were occurring and get back to the source on a timely basis, is the key to identifying the causes of reporting problems. It was well known that there were many problems with reporting transactions from the shop floor before the measurements were in place, but efforts to solve the reporting problems were unsuccessful. This failure can be attributed to trying to solve a poorly defined general problem, instead of several specific problems.

Another key element in improving performance was establishing accountability for reporting transactions. Before the performance measures were implemented, everyone was collectively responsible for reporting, but no one was specifically accountable for reporting any transactions. With the measures in place, accountability could be established down to the person reporting a transaction. Fear of making errors was never an issue, but talking to the people involved and clarifying instructions solved many problems. Getting input from the people involved about how to make reporting easier and more foolproof, eliminated many other problems.

REFERENCES
AND SUGGESTED READING

Amsden, Robert T., Butler, Howard E., and Amsden, Davida M. *SPC Simplified* (White Plains, NY: Quality Resources 1989).

Berry, Leonard L., Parasuraman, A., and Zeithaml, Valarie A., "The Service-Quality Puzzle," *Business Horizons* (September–October, 1988), 35–43.

Buzzotta, Victor R., "Workers Most Want What Many Leaders Least Provide." *Marriott Executive Memo* (November 7, 1997).

Camp, Robert C., *Benchmarking* (Milwaukee: ASQC Quality Press, 1989).

Case Study 23, *The Continental Insurance Companies* (Houston: American Productivity Center, 1983).

Chew, W. Bruce. "No-Nonsense Guide to Measuring Productivity." *Harvard Business Review* (Jan.–Feb. 1988), 110–118.

Cooper, Ken. *Conflict Management.* Audiotape (Fullerton, CA: TDM/McGraw-Hill, 1987).

Fuchsberg, Gilbert. "Quality Programs Show Shoddy Results." *The Wall Street Journal* (May 14, 1992), B1,B7.

Drucker, Peter. "Be Data Literate — Know What to Know." *The Wall Street Journal* (December 1, 1992), 10A.

Drucker, Peter. "The Information Executives Really Need." *Harvard Business Review* (January–February 1995), 54–62.

Frohman, Mark. "The Aimless Empowered." *Industry Week* (April 20, 1992), 64–66.

Garvin, David A. "Competing on the Eight Dimensions of Quality," *Harvard Business Review* (November–December 1987), pp. 101–109.

Hayslip, Warren R. "Measuring Customer Satisfaction in Business Markets." *Quality Progress* (April 1994), 83–87.

Herzberg, Frederick. "One More Time: How do You Motivate Employees?" *Business Classics: Fifteen Key Concepts for Managerial Success* (Boston: Harvard Business Review Publishing Division, 1975), 13–22.

Hindman, Stephen P. and Sviokla, John J., *Managing Top-Line Computer Applications*, Product #9-192-098, Boston: Harvard Business School, rev. 7-9-92.

Imberman, Woodruff, Ph.D. "Eliminating the Cost of Rejects." *Quality Digest* (October 1994), 55–59.

"Job-Cutting Medicine Fails to Remedy Productivity Ills at Many Companies," *The Wall Street Journal* (June 7, 1994), A2.

Kaplan, Robert S. and Norton, David P., *The Balanced Scorecard* (Boston: Harvard Business School Press, 1996).

Kaydos, Will. *Measuring, Managing, and Maximizing Performance* (Portland, OR: Productivity Press, 1991).

Kovach, Kenneth A."What Motivates Employees? Workers and Supervisors Give Different Answers." *Business Horizons* (September–October, 1987), 58–65.

Lynch, Richard L. and Cross, Kelvin F. *Measure Up!* (Cambridge, MA: Blackwell, 1991).

Marriotti, John L. *The Shape Shifters* (New York: Van Nostrand Reinhold, 1997).

Miller, Jeffrey G. and Vollmann, Thomas E., "The Hidden Factory." *Harvard Business Review* (September–October 1985), 142–150.

Myers, Scott M. "Who Are Your Motivated Workers?" Harvard Business Review (January–February 1964), 73–81.

Nelson, Bob. *1001 Ways to Reward Employees* (New York: Workman Publishing, 1994).

Schaffer, Robert H. and Thompson, Harvey A. "Successful Change Programs Start With Results." *Harvard Business Review* (January–February 1992), 80–89.

Taguchi, Genichi and Cushing, Don. "Robust Quality," *Harvard Business Review* (January–February 1990), 65–75.

Thor, Carl G. *The Productivity Measurement Handbook* (Portland, OR: Productivity Press, 1985).

Tufte, Edward R. *Visual Explanations: Images and Quantities, Evidence and Narrative* (Chesire, CT: Graphics Press, 1997).

What is Six Sigma? Document TI-29077 (Dallas: Texas Instruments, Inc., 1992).

ENDNOTES

[1] Gilbert Fuchsberg, "Quality Programs Show Shoddy Results," *The Wall Street Journal* (May 14, 1992), B1,B7.

[2] Mark Frohman, "The Aimless Empowered," *Industry Week* (April 20, 1992), 65.

[3] Ken Cooper, *Conflict Management*, audiotape, TDM/McGraw-Hill, 1987.

[4] Victor R. Buzzotta, "Workers Most Want What Many Leaders Least Provide," *Marriott Executive Memo*, November 7, 1997, 1.

[5] Frederick Herzberg, "One More Time: How do You Motivate Employees?" *Business Classics: Fifteen Key Concepts for Managerial Success*, 17 (Boston: Harvard Business Review Publishing Division, 1975).

[6] David A. Garvin, "Competing on the Eight Dimensions of Quality," *Harvard Business Review* (November–December 1987), 101–109.

[7] Leonard L. Berry, A. Parasuraman, and Valarie A. Zeithaml, "The Service-Quality Puzzle," *Business Horizons*, (September–October, 1988), 37.

[8] Carl G. Thor, *The Productivity Measurement Handbook* (Portland, OR: Productivity Press, 1985).

[9] For a good discussion of measuring productivity for practical applications, see: W. Bruce Chew, "No-Nonsense Guide to Measuring Productivity," *Harvard Business Review* (Jan.–Feb. 1988).

[10] Richard L. Lynch and Kelvin F. Cross, *Measure Up!* (Cambridge, MA: Blackwell, 1991), 65.

[11] Jim Robison, an ASQ Quality Committee expert estimates quality costs as percent of sales as follows: appraisal = 6 to 10%, prevention = 5 to 8%, internal failure = 10 to 12%, external failure = 8 to 12%. Source: reply to question over internet (www.tqnet.com). Other studies and sources provide similar figures.

[12] Jeffrey G. Miller and Thomas E. Vollmann, "The Hidden Factory," *Harvard Business Review* (September–October 1985), 144.

[13] Robert S. Kaplan and David P. Norton, *The Balanced Scorecard* (Boston: Harvard Business School Press, 1996), 97.

[14] Warren R. Hayslip, "Measuring Customer Satisfaction in Business Markets," *Quality Progress* (April 1994), 83–87. A discussion of how to improve surveys and survey techniques in business to business markets.

[15] Peter Drucker, "Be Data Literate — Know What to Know," *The Wall Street Journal* (December 1, 1992), 10A.

[16] Genichi Taguchi and Don Cushing, "Robust Quality," *Harvard Business Review* (January–February 1990), 65.

[17] Peter Drucker, "The Information Executives Really Need," *Harvard Business Review* (January–February 1995), 56.

[18] Robert H. Schaffer and Harvey A.Thompson, "Successful Change Programs Start With Results," *Harvard Business Review* (January–February 1992), 80–89.

[19] American Productivity Center, *Case Study 23, The Continental Insurance Companies* (Houston: American Productivity Center, 1983), 6.

[20] Robert T. Amsden, Howard E.Butler, and Davida M. Amsden, SPC Simplified (White Plains, NY: Quality Resources, 1989), 115–141.

[21] *What is Six Sigma?* Texas Instruments, Inc. (Dallas: Texas Instruments, Inc., 1992), 10.

[22] Woodruff Imberman, Ph.D., "Eliminating the Cost of Rejects," *Quality Digest* (October 1994), 55.

[23] Ibid.

[24] Robert C. Camp, *Benchmarking* (Milwaukee: ASQC Quality Press, 1989), 8.

[25] "Job-Cutting Medicine Fails to Remedy Productivity Ills at Many Companies," *The Wall Street Journal* (June 7, 1994), A2.

[26] Edward R. Tufte, *Visual Explanations: Images and Quantities, Evidence and Narrative* (Chesire, CT: Graphics Press, 1997).

[27] "What Motivates Employees? Workers and Supervisors Give Different Answers." *Business Horizons* (September–October, 1987), 61.

[28] Will Kaydos, *Measuring, Managing, and Maximizing Performance* (Portland, OR: Productivity Press, 1991), 157–160.

[29] Scott M. Myers, "Who Are Your Motivated Workers?" Harvard Business Review (January–February 1964), 73–81.

[30] Nelson, Bob, *1001 Ways to Reward Employees* (New York: Workman Publishing, 1994).

[31] John L. Marriotti, *The Shape Shifters* (New York: Van Nostrand Reinhold, 1997), 123.

INDEX